North Eastern Electric Stock 1904–2020

Its Design and Development

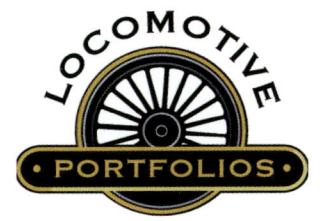

North Eastern Electric Stock 1904–2020

Its Design and Development

GRAEME GLEAVES

PEN & SWORD TRANSPORT

AN IMPRINT OF PEN & SWORD BOOKS LTD.
YORKSHIRE – PHILADELPHIA

First published in Great Britain in 2021 by
Pen and Sword Transport
An imprint of
Pen & Sword Books Ltd
Yorkshire - Philadelphia

Copyright © Graeme Gleaves, 2021

ISBN 978 1 52674 034 2

The right of Graeme Gleaves to be identified as Author of this work has been asserted by him in accordance with the Copyright, Designs and Patents Act 1988.

A CIP catalogue record for this book is available from the British Library.

All rights reserved. No part of this book may be reproduced or transmitted in any form or by any means, electronic or mechanical including photocopying, recording or by any information storage and retrieval system, without permission from the Publisher in writing.

Typeset in Palatino by SJmagic DESIGN SERVICES, India.
Printed and bound in India by Replika Press Pvt. Ltd.

Pen & Sword Books Ltd incorporates the Imprints of Pen & Sword Books Archaeology, Atlas, Aviation, Battleground, Discovery, Family History, History, Maritime, Military, Naval, Politics, Railways, Select, Transport, True Crime, Fiction, Frontline Books, Leo Cooper, Praetorian Press, Seaforth Publishing, Wharncliffe and White Owl.

For a complete list of Pen & Sword titles please contact

PEN & SWORD BOOKS LIMITED
47 Church Street, Barnsley, South Yorkshire, S70 2AS, England
E-mail: enquiries@pen-and-sword.co.uk
Website: www.pen-and-sword.co.uk

or

PEN AND SWORD BOOKS
1950 Lawrence Rd, Havertown, PA 19083, USA
E-mail: Uspen-and-sword@casematepublishers.com
Website: www.penandswordbooks.com

CONTENTS

	Introduction and Acknowledgements	6
Chapter 1	From Soot and Steel	9
Chapter 2	The North Eastern Railway	16
Chapter 3	The First North Eastern Electric Stock	27
Chapter 4	The Early Years of the Electric Services	36
Chapter 5	The Fire Replacement Stock	42
Chapter 6	The LNER Takes Over	45
Chapter 7	Old Trains, New Route	49
Chapter 8	New Trains, Old Routes	56
Chapter 9	Nationalisation, British Railways	67
Chapter 10	British Railways Electric Stock	79
Chapter 11	The End	89
Chapter 12	The Last Laugh	106
Chapter 13	Metrocars and Beyond	115
Chapter 14	Other North Eastern Railway Electrification Projects	130
Appendix	The Preserved Unit	146
	Bibliography	160
	Index	161

INTRODUCTION AND ACKNOWLEDGEMENTS

When sitting down to write this volume, after the procrastination is done and one puts on the writer's head, powers up the laptop and sits down to begin the process, a decision has been made, after very careful deliberation, about how best to approach this project. This process includes decisions on what exactly it is that I am trying to convey to the reader, and more importantly, who are the readers and what are their expectations, needs and requirements? This book is not the first volume to cover this topic and I suspect it certainly won't be the last. So what can I bring to the story that will make this a welcome addition to the ranks of other works by such well established and respected authors? I have been in the field of electric railway heritage for nearly twenty-five years now. I arrived here quite by accident but soon realised that I had found my niche, in not only writing about Britain's electric railways but in also saving for posterity some examples of the trains themselves (that is another book in its own right). Some of you may have read my previous work within this field. I have written other books and several articles for various publications over those intervening years. But each new work must be approached as exactly that, a fresh project.

My mission and my purpose is two-fold: firstly to avoid the easy option of simply putting out the same product as has gone before, but with a different name, and thus avoiding the trap of sticking to a simple tried and tested formula; and secondly to present my work in a way that is accessible and useful to as many readers as possible. I hope that in compiling this publication I have satisfied those two objectives and what has been undertaken achieves what I set out to do and that you, the reader, will benefit from the investment of your time in reading it.

Not being a resident of the North East and being born after the last of the original part of the network was switched off has not been a disadvantage to me in writing this book, such is the wealth of source material and the recollections of those who were there that, whilst researching and compiling this volume has certainly been a challenge, it is one that has brought me both great personal reward and sense of achievement. The journey started when I realised that the two-car unit parked on the Fullwell Curve alongside Strawberry Hill depot back in 1996 was not just another EPB unit but the last complete survivor of a system that at that time I knew next to nothing about. I knew it was called a Tyneside EPB but never understood why. I resolved to find out more. From that realisation I achieved not only self-education about the electric railways of the North East but also took on the challenge to get that two-car unit safely into preservation so it could be a physical representative of the story for many generations to come.

The story of the suburban electric railways around Newcastle and the North East of England is a story that has all the classic ingredients of any tale of industry and enterprise. In addition to the trains themselves there is the story of the ideas and imagination that gave rise to them, the personalities whose drive and ambition made those ideas become real, the science and technology, much of which was truly at the cutting edge for its time, that enabled it to not only happen but work day after day. There are also the innovations that were created and the lasting benefits they brought, some of which extended way beyond the original scope of the project.

Then there is the story of the people and the communities those trains served: the workers, the industry, the local commerce and the families. The story includes how their lives were transformed, enhanced and brought into a new age so far removed from that which had been the expectation for the preceding decades and how what was once a novelty became something so commonplace that it was accepted as the norm and even taken for granted by those who used it. Then there are the sadder sides to that story: the tragedies, the tales of neglect and the lack of vision that led to the rundown of the system. The two sides of the coin that starts with the inability to invest in and enhance what had been a truly worthwhile and beneficial venture that was then flipped to reveal a drive and ambition to correct that mistake and create something that once again put the area at the cutting edge of suburban transport and in the process setting a new standard that other cities would want to aspire to and emulate.

In this body of work I would be selling both you, the reader, and the story itself short by simply producing a list of dates, numbers and names accompanied with a selection of illustrative photographs and diagrams. I hope I have conveyed much more than that as I have aimed to do justice to both the story and the people who made it happen. As is the case with all my work I continue to add the disclaimer that my books are intended to be accessible to all, be you a railway historian, modeller, enthusiast, local historian or resident or even a casual reader; whatever your level of interest or reason for picking up this book then I hope there will be enough to hold your interest, inform and even educate you.

Finally, thanks must go to many people whose assistance, support and inspiration have made this project possible. I apologise if I have forgotten to mention everyone.

To start with there are my colleagues in the SERA 'Team Tyneside' project who are active at Shackerstone: Rob Davidson, Jacob Sparkes, Nick Hair, Andy Rowlands, Nick Evans and Dave Stretton. Some of them did the

Diagram of the all the lines that would be electrified by third rail over the period 1904–1967, along with the stations that would be open during all or part of that period. (Graeme Gleaves)

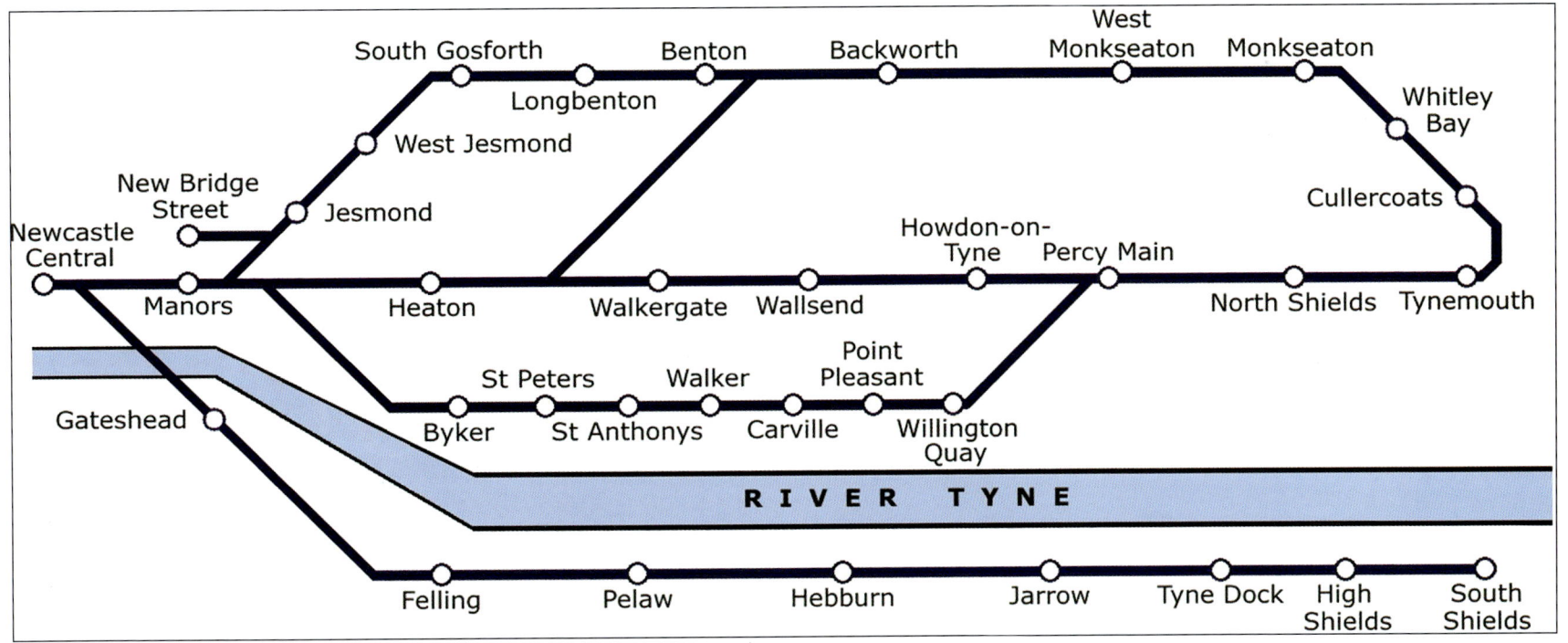

proof reading for the various chapters for me too. My thanks also goes out to Philip Champion who has been a supporter of the preservation project from the word go and has given me some insight of what it was like to be a schoolboy travelling on this network. Thanks also to Professor John Missenden for over two decades of support and guidance when needed. Anthony Coulls and Bob Gwynne of the National Railway Museum have also been of great help whenever I have called upon them and for that I am grateful. Thanks must also go to the staff at the Newcastle Central Library for their professionalism when I have gone there looking for reference material; they were always courteous and helpful and a credit to that fine institution. I must also recognise that it is my long-standing friend John Scott-Morgan who has made this book possible as it was all his idea in the first place. The project was devised over a pint in Woking and he has aided me further with photographs and useful contacts. I am further indebted to Laurie Kenward, John Atkinson and Robert Inns who provided information over the years, long before there was ever any talk of a book, and that information has been useful in compiling this volume. Thanks must also go to the contributors of images for this book. Most are credited with their images but mention must be made of: Richard Barber of the Armstrong Railway Photographic Trust, Brian Stephenson of Rail Archive Stephenson and Geoff Dowling of the Peter Shoesmith Archive.

On a personal note I wish to thank my family, Katharine, Floyd, Marina along with my children Sarah and Benjamin for their support, and finally the lovely Karen for her encouragement to get this project finished. You may have no idea what this book is about and will probably never read it but, darling, you've left your mark on its every page.

Graeme Gleaves
Slough
Twitter: @MrLangelo

Chapter 1
FROM SOOT AND STEEL

The north east of England is one of the most important areas of Britain; throughout modern history the region has played more than a passing role in the economic and industrial development of our nation. At the heart of the region is the great city of Newcastle-upon-Tyne, where the Romans built the first bridge across the river here some time around AD 122, and which marked the northern limit of the Empire of Emperor Hadrian. That same year, to the east of the settlement where the north bank of the river meets the North Sea he ordered the construction of a great wall that would stretch across northern England to the coast of the Irish Sea. Wallsend gets its name from this structure as it was literally the eastern end of Hadrian's Wall. The Romans would also be responsible for first exploiting the natural resource that would bring wealth and prosperity to the

A drawing of West Moor Pit, Killingworth. This was typical of many pitheads of the area during the early nineteenth century. The buildings and the winding gear are clearly visible, along with the chaldrons loaded with coal and being pushed by hand. (Author's Collection)

An illustration from an early nineteenth century French book depicting a coal chaldron in the north east of England. There is a bit of artistic licence here given that the wagon is running on rails but the wheels appear to have no flanges and the gentleman riding the wagon appears to be a little overdressed for his job. Nonetheless it does give an idea of one of the earliest applications of railways in the area. (Author's Collection)

region: the seams of good quality coal. Whilst it would be hundreds of years until the mechanised industrial mine was developed, the working of coal in the region continued on a small scale through the Middle Ages. It was not until a fuel crisis hit Elizabethan England in the last half of the sixteenth century that coal began to be more intensively exploited. Up until this time wood was the fuel that ran Britain; it also built the nation's houses and ships, but it was running out. Trees were being felled faster than they could be replaced.

Coal from the North East was the answer to the crisis. By 1630 over 400,000 tons of coal were shipped from the River Tyne and by the end of the seventeenth century that figure had reached over half a million tons. The abundance of coal in the region was a source of prosperity for a select few, and a chance of employment for the rest. It also prompted the development of other industry in the area that needed coal supplies to function. But it was the advent of the Industrial Revolution that would transform the north east of England at a rate of change that was staggering. The Elizabethans had shed their reliance on wood and popularised the use of coal as a fuel. This paved the way a century and a half later for the Industrial Revolution to show just how useful coal was, and to do this the country was going to need a lot of it, along with the means to transport it the length and breadth of the country. The North East was going to be the cradle of the new technology of railways.

An enthusiastic engineer working at the Killingworth Colliery just to the north of Newcastle, inspired by the work of early engineers like Richard Trevithick, persuaded his employers that a steam locomotive would aid productivity in transporting mined coal. His name was George Stephenson and he would go on to build some of the first commercially successful locomotives in Britain and, along with his son Robert, would be responsible for the creation of the early railway network that would go on to to change the world. Robert Stephenson's premises were in Newcastle-upon-Tyne and it was here that *Locomotion No. 1* was built for the Stockton and Darlington Railway, followed in 1829 by the *Rocket* which won the Rainhill Trials and thus paved the way for steam locomotive design for the next 125 years. Robert was also the engineer who built the High Level Bridge across the River Tyne which opened in 1849.

The spread of railways across Britain was meteoric, and within twenty years there was a network of thousands of miles of track linking all the major cities. Newcastle had a grand station built that was designed by John Dobson and opened by Queen Victoria and Prince Albert in 1850. Newcastle Central station was the largest building of the nineteenth century to be constructed in the city, and as the railways spread became a site of intense activity. Lines radiated out from the station to feed into the suburbs of the city and the industrial areas of the locale as well as the trunk routes to London and north to Scotland. Dotted amongst these lines were the wagonways of the collieries that were bringing coal from the pits to the staithes on the river for loading onto boats that would carry it to other parts of the country and also for export to

George Stephenson (1781–1848), a man from the North East of England who played a huge role in the development of not only railways but the modern world. (Author's Collection)

Robert Stephenson (1803–1859), George's only son and an equally gifted engineer. He was responsible, amongst other things, for establishing locomotive building in the North East and the design and construction of the High Level Bridge over the River Tyne. (Author's Collection)

Europe and beyond. Coal staithes were elevated platforms where the coal was unloaded from the wagons and into the ship below. Britain had established herself during the Industrial Revolution as the largest producer of coal in the world. Coal needed ships and it was here that Tyneside had another source of industry. On both the north and south banks of the river shipbuilding yards were established that could build anything from a coal barge to a naval cruiser. The Swan brothers, Charles and Henry, were two natives of Tyneside who excelled in the world of shipbuilding. Charles took over the yard of Coulson, Cooke and Co. in 1874 and changed its name to C. S. Swan & Co. Charles Swan was fatally injured five years later when he fell into the rotating paddles of a steamer whilst crossing the English Channel on the way home from a business trip to Russia. His widow went into a business partnership with George Hunter, a shipbuilder from Sunderland, and thus the name Swan Hunter was born, a brand that still survives on Tyneside but has not built any vessels since 2006. The other Swan brother, Henry, went on to take over the running of the Mitchell shipyard in Walker. He specialised in oil tankers and during the last two decades of the nineteenth century he was responsible for building over half the world's oil carrying vessels.

The railway and shipbuilders generated other industries in the area where there was a market for

The early stages of the construction of the ocean liner RMS *Mauretania* at Swan Hunter's yard at Wallsend. Ship building had become a key element of the area's economy by the start of the twentieth century. (Author's Collection)

Another view of the *Mauretania* in 1906 shortly before being launched. At the time she was the largest ship in the world and hundreds of Geordies were responsible for making her the pride of the Cunard fleet. (Author's Collection)

their products. Boiler makers, crane builders, munitions makers and of course steel manufacturers all set up shop on the banks of the Tyne. This resulted in a staggering growth in the population. The 1851 Census shows a population of 87,000. By 1901 this figure had risen to over 225,000.

Technical innovation continued on Tyneside well into the latter half of the nineteenth century. The Newcastle Literary and Philosophical Society had a list of members who read like a who's who of engineering excellence of the time. At a meeting of the 'Lit and Phil' in 1878 Joseph Swan demonstrated his new invention to an astonished audience: the electric light bulb. Swan had placed a glowing carbon filament inside a glass vacuum tube, a revelation to a world that was at the time reliant on candles and oil lamps for illumination. Over the Atlantic at the exact same time Thomas Edison would do the self same thing, but in Great Britain the lightbulb was invented by a Tynesider and not an American. Edison would share a patent with Swan but unfortunately history would really only remember the former. In 1881 Mosley Street in Newcastle became the first street in the world to be lit by electricity.

Another innovation to originate from the area was the steam generator that was patented in 1884 by Charles Parsons. He set up his own firm in Heaton to develop the technology and it gave rise to his steam turbine a few years later that saw great application in ships. The steam generator found another

Contemporary drawing from around the time of the opening of Newcastle Central station. There is much activity on the forecourt, with workmen busying themselves and the coming and going of many carts and carriages all drawn by horses. (Author's Collection)

This illustration shows the new High Level Bridge over the River Tyne along with the older Tyne Bridge in the foreground, which would be removed in 1876. (Author's Collection)

application as the means by which electricity could be produced in far greater quantities than before. The first power station in the world was opened in late 1881 in Godalming, Surrey, and was driven by a water wheel. This was followed in January 1882 by the world's first coal-fired power station, which was located on Holborn Viaduct in London and opened by Thomas Edison's Electric Light Company. This was still a small scale affair, but in time it was the steam generator that was to be the key to future widespread power generation the world over.

Parsons established The Newcastle and District Electric Lighting Company (DISCo) in 1899 and a year later Tyneside got its first power station at Forth Banks, very close to the central station. It used Parsons' turbo alternators which were fed steam by three Lancashire boilers; the alternators delivered electricity at 1,000 volts for distribution to smaller substations. A rival firm, the Newcastle-upon-Tyne Electric Supply Company (NESCo) was founded by local industrialist John Theodore Mertz and opened Pardon Dene power station in the east end of the city the same year. It burned 2,700 tons of coal each year to feed the boilers that drove the steam engine which turned the alternators to put out power at 2,000 volts; this was fed to local transformers and from there went at 100 volts to the consumers. Both concerns were supplying power that was mainly for lighting, be it streets, houses or commercial premises. NESCo was located in the east end of the city while DISCo supplied the areas to the west of Grainger Street, and as can be imagined there was fierce rivalry between the two concerns. Despite this it did put Newcastle at the forefront of power generation and it was only a matter of time before a new and bigger power station would be built to satisfy the demands of the industrialised areas around the city.

NESCo's founder John Mertz had moved to Newcastle from Manchester; he was an Anglo-German Quaker who trained as an industrial chemist. It was from here he developed a keen interest in the emerging science of electricity. This enthusiasm was shared by his eldest son, Charles Hesterman Mertz, who served his apprenticeship at NESCo before he did a spell as secretary for the Cork Electric Tramways and Lighting Company in Ireland before returning to work for his father at NESCo again at the Pardon Dene power station in 1892. Both John and Charles saw the possibilities electricity offered beyond just lighting and were advocates of its greater use in industry. Charles had set up a consultancy firm in 1899 to promote new installations, no doubt drawing upon the work of his father and his own experience in Ireland. That consultancy was enhanced in 1902 by the arrival of William McLellan, a Scottish engineer who had met Charles whilst working in Ireland. The consultancy firm of Mertz & McLellan was to have a profound effect on the development of electric transport in the north east of England but their first major project was the re-design of the Neptune Bank power station at Wallsend on Tyneside. The redesigned facility was commissioned in 1901 and was the first power station in the world to generate three phase electricity; all previous stations had generated at a single phase. Neptune Bank had been opened by the Walker and Wallsend Union Gas Company in 1899 but NESCo took it over and upgraded the equipment before the re-commissioning in June 1901. The station could produce 2,800 kilowatts of power at 5,500 volts. The output was shortly to be increased to 3,000 kW by employing turbo alternators designed by Charles Parsons.

In 1901 Newcastle Corporation Tramways opened the first section of their electrified network. Trams had first appeared in and around Newcastle in 1879 with the small network that had been built between Newcastle and Gosforth High Street using horse drawn operation. Newcastle was slightly late in adopting electric trams as both Gateshead and Middlesbrough had their first routes before 1900. The Corporation Tramways undertaking was to be far grander as it was to be operated with electric traction, with its own power house built at Manors to supply the current needed for the overhead trolley wires. A deal was struck to link up with the Tyneside Tramways and Tramroads Company (TT&TC) who were constructing a tramway that would run from Gosforth through Wallsend to North Shields; the trams of this company were also to be worked by electric traction. The Chairman of the TT&TC was none other than John Mertz and he had his son Charles engaged as a consultant on the scheme to advise on the use of DC current and its supply from both the Manors power house and from NESCo.

A third tramway running from North Shields via Tynemouth to the coastal destination of Whitley Bay converted its operation from horse

An illustration depicting the arrival of Queen Victoria and Prince Albert by train at Newcastle Central station in 1850 for the purpose of formally opening the station. It was the second rail trip to the area by the monarch in as many years as she had opened the High Level Bridge the year before. (Author's Collection)

drawn trams to electric ones in March 1901. This was operated by the Tynemouth District Tramways Ltd and was a separate concern with a different track gauge of 3 foot 6 inches, making through running with the other tram companies impossible. An interchange was provided between their trams and those of the TT&TC at New Quay in North Shields, giving a reasonable connection for the electric tram that would take passengers directly onwards to the city.

The areas of Wallsend, North Shields, Manors and Gosforth that were now served by electric tram cars were populated residential districts with high numbers of patrons that needed conveyance to their places of work, either in the city or the industrial areas on its outskirts. The Tynemouth and Whitley Bay areas represented both work and leisure travel, with the latter destination being very popular at weekends and holidays when Tyneside residents headed to the coast in great numbers to catch the sea air or enjoy the sandy beach. The electric trams now offered a frequent service at a cheap fare on a modern form of transport; this was good for the people of Tyneside but bad for the operators of the traditional railway services in those areas. The North Eastern Railway (NER), who had enjoyed an unchallenged position for patronage now found a major source of income was being taken from them. There needed to be a fight back and thankfully the man to best advise them how to do it was on their doorstep. The NER decided to call upon the services of Charles Hesterman Merz.

Chapter 2
THE NORTH EASTERN RAILWAY

The north east of England had become dotted with wagonways long before anyone thought to put a fare paying passenger on a train. The movement of coal from the pits to the staithes where it could be transported onward for industrial and domestic use was the sole reason for their existence. These wagonways were operated in the first instance by either horses or manpower, then the steam winding engine and rope haulage became viable. This was followed in most cases by the steam locomotive in the second quarter of the nineteenth century. Passenger railways were a secondary consideration after the movement of freight for the early railway pioneers, but given time the idea did eventually catch on. The Victorians were never shy to exploit a chance to turn a profit.

A postcard from the late nineteenth century showing the famous diamond crossings at the north end of Newcastle Central station. The suburban platforms for coastal trains are on the right and a train of pre-electric steam hauled stock is in one of the suburban platforms. (William Smith)

Tynemouth sands at the start of the twentieth century. Having this pleasant stretch of coast within easy reach of the city generated much revenue for the railway company. (Author's Collection)

Tynemouth station pre-electrification. A four wheeled carriage can be seen in one of the bay roads on the right. The size and layout of the station gives an indication of the volume of traffic the coastal routes north of the river generated. (Author's Collection)

A view of the interior of Tynemouth station, famed for its floral displays. The NER porter stands ready with his trolley whilst the news vendor has copies of the *Chronicle* for sale. The layout and poses of the people tend to indicate this was a staged picture and not an impromptu scene. (Author's Collection)

The first passenger railway reached Newcastle in 1839 when the line from Carlisle extended to a terminus at Shot Tower. Prior to this, for three years passengers had to leave the train at Derwenthaugh and be taken over the river by ferry. A second route opened up the same year when the Newcastle and North Shields Railway opened its railway from a terminus at Carliol Square to North Shields Quay. The line would be extended to Tynemouth in March 1847 by which time the Newcastle and North Shields Railway Company had been absorbed into a larger concern by the name of the Newcastle and Berwick Railway. That company had opened its first section of line north from Heaton to Morpeth in March 1847 and in July of that year had commenced running to the Carliol Square terminus. The Newcastle and Berwick Railway had become the York, Newcastle and Berwick Railway Company (YN&BR) in late in 1847 when it absorbed the York and Newcastle Railway. This growing empire was run by the infamous 'Railway King' George Hudson, who promoted multiple railway companies and made a fortune during the railway mania of the 1840s, only to lose it all through financial malpractice that not only forced him to lose his seat in Parliament but to also flee the country in 1849 to avoid arrest. The YN&BR carried on without him, completing the Central Station in Newcastle in 1850 and continuing to expand its area of operations by either promoting new lines or absorbing smaller companies. In 1854 it changed its name to the North Eastern Railway (NER) after amalgamations with other companies. In 1874 it absorbed another company that had built lines feeding into Newcastle-upon-Tyne, the Blyth and Tyne Railway, who served their own terminus in Newcastle at New Bridge Street with a line that ran south from Blyth via Seghill, Monseaton, Backworth and Gosforth. They had connected to the YN&BR at Tynemouth so the absorption of the Blyth & Tyne by the North Eastern Railway was inevitable given that the latter company had a near monopoly of the routes within the counties of Durham & Northumberland. By the close of the nineteenth century the NER had an empire that was based in offices at York and served nearly every line from north of Doncaster to the Scottish border at Berwick-upon-Tweed; they even had some route miles of their own in Scotland and have remained the only English based company ever to do so.

On Tyneside they ran an hourly service between New Bridge Street and Newcastle Central via the former Blyth and Tyne and Newcastle & North Shields routes; a loop line had been added to the latter in May 1879 between Manors and Percy Main which branched off and followed the contours of the north bank of the River Tyne. It's known as the Riverside branch, despite it being a loop line and not a true branch line at all. The Riverside route served the heavily industrialised areas and was a source of traffic for both the

The crowds enjoying the sands at Whitley Bay. The trip to the seaside was popularised during Victorian times and became a highly accessible recreation. Nearly all these people would have been ferried to and from the location by train. (Author's Collection)

Above left: **The station** clock says it is just after half noon but Central Station is still bustling with locomotives, rolling stock and staff, not to mention passengers and their luggage. Platform 8 is a through platform used by services heading to Scotland. The suburban platforms are behind the building and trains to the centre left of the image. (Author's Collection)

Above right: **The forecourt** of Central station has changed considerably since the drawing in the last chapter and is now Neville Street. Not only carriages but horse drawn trams ferry people to the station in this late nineteenth century drawing. (Author's Collection)

Right: **Horse trams** of the Newcastle Tramways pictured in Gosforth High Street at the turn of the nineteenth century. (Newcastle Central Libraries)

workers going to and from their employment as well as goods traffic serving the factories, plants, shipyards and engineering facilities they worked in. The remainder of the route, which formed a circular elongated 'horseshoe', was made up of areas where the expanded population created by the effects of the industrial revolution now lived. These were working men employed in industry or white collar workers who worked in the commercial districts in the city. Then there were the shop workers, the market traders and also those who wanted to escape their suburban homes at weekends and head for the coast at Whitley Bay, Tynemouth or Cullercoats. All in, this area was a rich source of regular income for the NER, and the lines were responsible for 9.8 million passenger journey in 1900/1901.

So the NER did not take kindly to the electric trams that started running in the spring of 1901. In the year that followed, their receipts for the area dropped by 57 per cent. When the through running of trams between Newcastle and North Shields was scheduled to begin in the near future the NER would be unable to sustain a further loss of traffic. In February 1902 the board of the NER discussed the issue in response to a memorandum from their Chairman, George Gibb, which suggested the North Eastern Railway meet the challenge head on by electrifying their railway. They called upon the services of Mertz & McLellan to advise them of the extent of the problem and their possible options on how best to meet the challenge of the electric tramcar.

The report Mertz & McLellan produced set out in detail where the NER was losing out. The trams were run at frequent intervals and at a fare that was affordable to the populace. The tram stops were usually within five minutes' walking distance from either the passenger's residence or their intended destination, often both. Because they were so frequent there was never a need to consult a timetable and next to none of the tram's patrons even possessed or looked at one. On the contrary the NER trains ran at an interval of only one an hour with a few extras at peak times. The equipment was old and did not present a 'modern' travelling environment. The report did go on to say that the trains did posses a number of inherent advantages over the tramcar: they were faster over longer distances due to the less frequent stops and the ability to use heavier, more powerful traction. The trains had a much higher capacity than the trams so could carry far more people per route mile and thus in turn generate greater income.

The new order! A fully loaded electric double deck tram has replaced the horse drawn ones on the Gosforth route and now offers a cheaper, more frequent alternative to the North Eastern Railway. (Newcastle Central Libraries)

Above left: **The staff** of the Tyneside Tramways & Tramroads Company proudly pose with their new sleek electric tram that will enter service on the Newcastle to North Shields route, offering the NER real competition for trade. (Newcastle Central Libraries)

Above right: **Construction of** new tramway routes in Newcastle city centre in 1903. (Newcastle Central Libraries)

A retouched postcard of Grainger Street, Newcastle looking toward the Grey's Monument. Hand carts, horse and cart and of course the new electric tram are very much in evidence. (Author's Collection)

The report pointed out that not all of the patronage of the tram services had been taken from the railways, as much of it was new custom who were attracted by the convenience of both the service and the cost. Mertz & McLellan's report gave the following recommendations to the NER.

- Electrify the suburban routes around Newcastle and the coast.
- Introduce a frequent service of electric trains.
- Offer a service that would be flexible to the demands of traffic, with longer trains during busy periods and increased frequencies.
- Accept that trams would deal with most of the short distance local traffic and that the NER should aim to monopolise those journeys of more than two miles.

Of course the cynic may say that by advocating electrification Charles Mertz was simply trying to drum up business for his father's NESCo concern, but that is simply not the case. The report submitted to the NER made perfect sense, and the points made about the needs of local urban and suburban transport are as valid today as they were over 116 years ago. Whilst NESCo were going to benefit from any NER electrification, that is an accident of history and merely indicates that the Mertz family were in the right place at the right time. Charles Mertz would go on many times to prove to be a visionary in the area of industrial and domestic power supply. He is considered by many to be the prime mover behind the creation of the national grid in Britain and was awarded the Faraday Medal for his work in 1931. He was working up until his untimely death in 1940 when a German bomb fell on his house in Kensington during the blitz, killing both Charles and his two children. He is still honoured on Tyneside by a building named Mertz Court on the Newcastle University campus. The offices he and William McLellan used for their initial consultancy business in Gosforth have a blue plaque recording their former use.

At the end of 1902 the NER Board read and debated the report, accepted the findings and agreed to proceed with a comprehensive electrification scheme. Drawing on best practice from within the UK was not easy, as at this time very few lines had been electrified: there were the deep level tube lines of the City and South London and the London and South Western Railway's Waterloo and City lines, along with the open air elevated route of the Liverpool Overhead Railway. Other schemes were in the planning stage but it was the tram routes that had been quickest to exploit electric traction. However, the needs of a tramway were not applicable to the needs of a larger, heavier and more demanding suburban railway where the permanent way had to be shared with other traffic such as express passenger trains, goods trains and local traffic over other non-electrified routes. One country that was already ahead of Britain in the introduction of electrification of its suburban routes was America and it was from here that the NER would draw most inspiration in shaping their Tyneside electrification scheme.

Charles Hesterman Mertz (1874–1940), the consulting engineer who brought electricity to the transport network on Tyneside. (Author's Collection)

The city of Chicago was where the greatest influence on future British electrification could be found. The South Side Elevated Railroad had opened in 1892 and served areas of downtown Chicago to Jackson Park, which was reached in 1893 in time for the World's Columbian Exposition Fair that was being held there. The line

Wilson Worsdell (1850–1920). The Chief Mechanical Engineer of the North Eastern Railway from 1890 until 1910, he succeeded his brother in the position and presided over the electrification project for the North Tyneside lines. (The Engineer)

A six car multiple unit on the Chicago South Side Elevated Railway in 1903. The outline and operation of these units was a blueprint for not only future EMUs in America but also the first generation of electric stock in the UK, including those of the North Eastern Railway. (Author's Collection)

was on an elevated structure and steam-operated by 28 ton Baldwin built 0-4-4 tank engines hauling carriages, and the system ran for twenty-four hours of each day. The line got into severe financial difficulties only a few years after opening and was sold in 1896 to a new company who planned to rescue the operation and electrify it using a third rail system that had found favour on one of Chicago's other routes, the Metropolitan West Side Elevated railroad. In 1897 the line converted not only to 600v DC third rail operation but also employed a completely new type of electric train, one where there was no locomotive at the front but a number of motorised coaches along its length, together with unpowered trailer cars. This was the world's first electric multiple unit and had been the creation of Mr Frank Julian Sprague. A single driver controlled all the motors on the train from the lead driving position, by a system of control lines and relays that ran the length of the train, with electrical connections between the coaches. This system would become the standard for electric railways the world over as it offered the advantages of spreading the weight of the traction equipment over the whole train rather than into a single locomotive. Furthermore, the train could enter and leave the terminal in a fraction of the time it took to change locomotives. Both Charles Mertz and the NER's Assistant Chief Mechanical Engineer, Vincent Raven, visited America. How much they saw of the Chicago system is not recorded, but it would be a near duplicate that would be built on Tyneside, right down to the choice of conductor rail and the voltage that was fed to it.

Previous electrifications in Britain had followed one of two practices, one where the conductor rail is placed between the running rails and the negative return effected through the running rails, and the other where the conductor rail is outside of the track gauge with the negative return effected by an additional conductor rail placed between the running rails, this latter system proving highly suitable for underground railways where the insulation offered by a dedicated return rail would offset the problem of earth leakage in a subterranean environment. The system is still used today by the London Underground.

The NER were the first to adopt the Chicago system whereby the conductor rail, fed at 600v DC is placed outside of the running rails and frequently changes sides to even out the wear on the train-mounted collector shoes that make contact with the top surface of the live conductor and transfer the power to the train's positive bus line. The current is then used by the train's motors and other electrical equipment after which the negative return current is fed through the wheels to the running rails, where the circuit with the point of supply can be completed. This system still finds extensive use on lines in the south of England where it is known as the 'Southern Electric', despite the fact it was first introduced to the country in the North East and is an American design. Common to all systems was the placing of the elevated conductor rail on 'pots' made of porcelain, or occasionally glass, to insulate it from the ground.

The NER railway broke with another tradition of previous electrification projects on these shores by not having their own power station. Instead they bought their electricity from NESCo, who were in the process of building a huge new power station on the north bank of the river near Wallsend to be known as Carville Power Station. The station was not built purely to electrify the NER, although they would be a major customer. NESCo had eyes on the supply of power to the shipyards and mines of the area as the use of electricity in industry was on the rise. Carville was built on a fifteen acre site and was designed by Mertz & McLellan. It was equipped with four turbo-alternators produced by Charles Parsons' works and which were fed by ten Babcock & Wilcox marine boilers. The initial output was up to 10,000 kW, at a three phase output of 6,000v which was fed into NESCo's distribution network. The station would undergo many upgrades and extensions over the next few years and would set the standard for power station design the world over for decades to come.

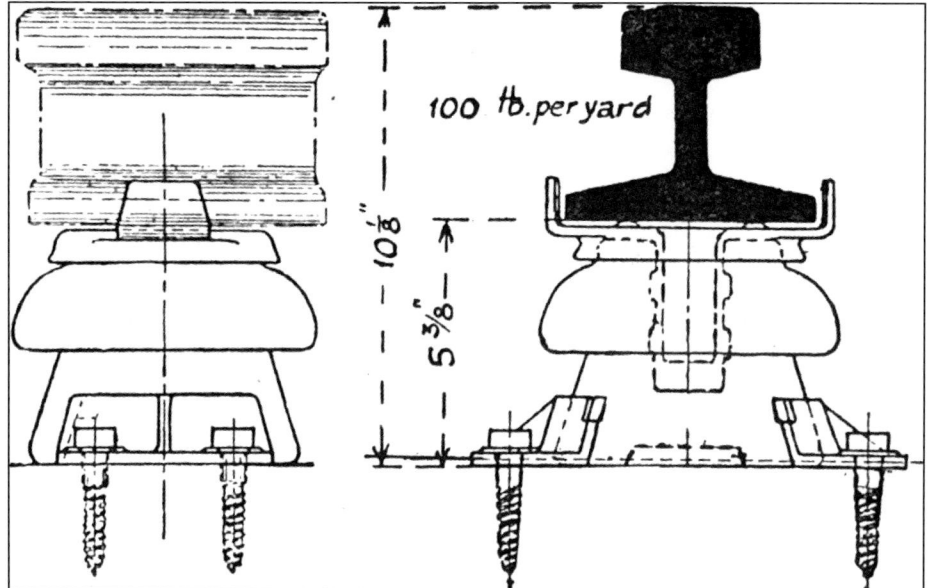

Side and cross section diagrams of the third rail and porcelain insulators used by the North Eastern Railway, sometimes referred to as the Vignoles system due to the use of flat bottomed conductor rail. Charles Vignoles was a railway engineer who championed the use of flat bottomed running rails yet his name ended up being associated with flat bottom conductor rails. (Author's Collection)

One of the combined steam turbine/alternator sets from Carville power station that supplied power to the NER's third rail network. (The Engineer)

On the ground, work began in early 1903 to equip the NER's lines for electric operation. The plan included both the coastal circle from New Bridge Street around to Newcastle Central along with the Riverside branch, and the Ponteland branch line that was under construction from a junction north of South Gosforth to head north west and serve communities on the outskirts of the industrialised and commercial area. The Ponteland line would open for traffic in June 1905 but failed to reach the expected levels of passenger traffic and thus the electrification of the line never happened during the NER's tenure. The line would one day play a role in Tyneside's electric railways, but that would be many decades into the future.

Numerous firms were contracted to supply the equipment. Siemens Bros. & Co. Ltd supplied the high tension cabling to connect the substations to the NESCo grid, whilst the equipment inside the substations that would convert the supplied voltage down to the 600v DC required for traction was supplied by British Westinghouse & Electric Manufacturing Company, set up by the American George Westinghouse who opened a plant in Trafford Park, Manchester in 1899 and was also heavily involved in the electrification of the Mersey Railway that was completed in 1903. British Thompson-Houston Company Ltd got the contract for equipping the electric cars that would provide the service, more of which in the next chapter. The total cost of these three contracts was just over £300,000; a not inconsiderable sum of money to even a large company such as the NER. This was a something of a gamble but they must have had faith in the opinions of Mertz & McLellan to continue to keep them on retention for further reports on electrification at the sum of £10,000. The NER stipulated that completion of the electrification work be made in sections during 1903 with an estimated commencement for the start of electric services set for early 1904 once all testing and staff training had been completed and the whole operation had been approved by the Board of Trade.

Chapter 3
THE FIRST NORTH EASTERN ELECTRIC STOCK

The date of the public launch of the electric passenger service between New Bridge Street and Benton was Tuesday 29 March 1904. At around noon a ceremonial 'first' service for local dignitaries, NER officials and some of the press was started from New Bridge Street by Lord Ridley, the then Chairman of the NER and a former Home Secretary. The special ran to Benton and back and thus ushered in the electric age on Tyneside. The service commenced for the public in each direction with the 12:50pm departure from New Bridge Street and the 1:03pm return service from Benton. It is now appropriate to look at the very first batch of stock provided by the NER for the opening of the newly electrified routes.

The initial batch was for ninety cars, all but two of which were passenger carrying. They were constructed through 1903 and added to NER stock in 1904. It is fair to say that the original fleet of cars had a distinctly 'American' look to them. This was not unique to the NER as electric units being built around the same time for the Lancashire and Yorkshire, Metropolitan, and London's deep level tube lines shared many similar characteristics and would all have not looked out of place in a wild west movie. Indeed the American term 'cars' was used to describe individual vehicles instead of the more traditionally British term of carriages. The first batch of cars built for the NER can be divided into eight different types, detailed in the table.

All the vehicles were constructed at the NER's York carriage works with the bogies, motors and electrical equipment supplied by an outside contractor with previous experience of such work, namely the British Thomson-Houston Co. Ltd. The original contract had been for 100 vehicles but the order was scaled back to these initial 90 with an

Table 1: The first batch of electric cars built for the NER.

Diagram No.	Vehicle Type	Number Built	Car Number Range	Notes
90	Driving Motor Luggage First	24	3239–3262	Single driving end with luggage compt.
91	Driving Motor Third	24	3213–3235	Single driving end.
92	Trailer Third	27	3180–3206	
96	Driving Motor Third	1	3236	Double driving end, half glazed platform facing doors and no gates.
97	Driving Motor Third	2	3268–3269	Double driving end, shorter longitudinal seats and more reversible transverse seats.
98	Driving Motor Luggage Third	5	3237–3238 3263–3265	Double driving end with luggage compt.
99	Driving Trailer Third	5	3207–3211	Double driving end, as per Diagram 96
100	Motor Luggage Van	2	3266–3267	No passenger accommodation

option to add more as demand for the traffic dictated. As delivered, the standard three coach train would have comprised a Diagram 90 Driving Motor Luggage First and a Diagram 91 Driving Motor Third with a single Diagram 92 Trailer Third sandwiched between them. The arrangement of the cars was such that this was only one possible combination as vehicles could be added to or removed from a formation relatively easily to make a longer or shorter train depending on the needs of the traffic. All that was required was a driving cab at each end of the formation for it all to work perfectly.

The first three diagrams, namely 90, 91 and 92, had the passenger access from the platform by way of hinged iron gates, not dissimilar to something that would grace the front step of a respectable town house. This led onto a vestibule area where a glazed door led into the spacious open saloon. One has to imagine how this layout would have appeared to the seasoned commuters who had become accustomed to the gaslit full width compartments of the steam hauled stock the new electrics were to replace. Entering the open saloon, be it first- or third-class, the passenger would be offered a choice of longitudinal or transverse seating. In first class this was upholstered in leather, whilst in third class rattan covering was used for the seating. Suspended from the centreline of the clerestory roof ceiling in both classes were glass shades for the electric lighting. Electric heating kept the interior warm during colder weather and ventilation was in the form torpedo vents fitted either side of the clerestory roof, with passenger operated brass ventilator openings at head height inside the saloon. Leather straps for standing passengers were provided in all vehicles as were net luggage racks

This picture is believed to show an NER electric train at Tynemouth station on the very first day of electric services extending to this point, on 21 June 1904. The large number of people in the cab and the well turned out crowd on the platform certainly indicate this was not an ordinary day on the network. (Newcastle Central Libraries)

The interior of a 1904-built NER electric unit. This is also believed to be the inaugural service and the fact that this is a third-class car and the passengers are very well dressed tends to support the claim. (Newcastle Central Libraries)

longitudinally above the windows, regardless of which direction the seating faced. Full sized side light windows were fitted to all vehicles with narrow top-lights on the outer edge of the clerestory roof, none of which opened and which gave rise to complaints of poor ventilation. The carriage body was constructed over an all steel underframe, 55 feet long with wooden framing to which the body and roof sections were attached. The body length came in at 56 feet and 6 inches and had an overall width of 9 feet. The lower portion of the bodies was of vertical matchboard construction which was not something normally associated with suburban rolling stock and gave these vehicles an air of extra elegance. This was enhanced by the choice of livery which was different to that applied to other NER stock. It featured a red lower body side with cream upper; the underframe and bogies were finished in black. The company name was carried on the sides of the cars at cantrail level and car numbers and lettering was applied in gold to each vehicle halfway down the red matchboard area.

The driving cabs were full width with the driving position on the left-hand side of the cab. Originally the cabs were provided with a full size driver's window but these were changed, more of which later. A hinged door was provided in the centre of the cab for crew use only, and above the cantrail were three lamps used to indicate the route of the train. Additionally a glass plaque was carried in the front stating the destination and a brass whistle was fitted on the roof over the driving position, operated by compressed air from a valve in the driving cab. A Driving Motor Coach normally weighed in the region of 30 tons and an unmotorised Trailer Car was considerably lighter at 21 tons. The first complete three-car train was out-shopped from York in September 1903. It was personally inspected by the NER's Chief Mechanical Engineer Mr Wilson Worsdell, his assistant Vincent Raven, General Manager George Gibb, and other officers of the company at York station. The set was towed as trailing load by a steam locomotive to evaluate its ride and structural integrity over two days on trips from York to Knaresborough and Ripon, before being hauled north to Heaton car sheds on 26 September so it could be electrically tested on the section of line that had been energised between Percy Main and Carville. The first test run was carried out on 27 September.

All passenger vehicles that were equipped with motors were fitted with two G.E. Type 66 motors of 125hp. These were mounted on the same bogie, under the driving cab end of a single cab vehicle. Current collection was by collector shoes fitted to the driving vehicles on each bogie. These were the gravity type that were suspended from the leading end of a wooden beam on the side of each bogie. Heavy duty cabling was provided to feed the electrical supply from the collector shoes to the train's main power line, from where it could be distributed to the control, motors, lighting, heating, air compressors for the Westinghouse air brakes and the air sanding fitted to the leading bogies. The collector shoes were relocated to a central position on their mounting beams from an early stage as it was found this improved the contact area when the bogies pitched on curved track.

Couplings were unique on the Tyneside electric stock, being of the 'Cowhead' type. This coupling was found nowhere else on Britain's railways but was standard on three of the four generations of Tyneside electric units. It was not an automatic coupler but did have a drag box behind it and under the end of the car's frame that alleviated the need for separate buffers to be fitted. The exceptions to this were the two non-passenger fitted luggage vans that had Cowhead couplers and also side buffers, like steam hauled stock, which were required as they became frequently required to haul vans and other coaching stock.

The initial testing concluded on 8 November when the trial train was returned to York for rectification work. Construction of

A service to New Bridge Street is making ready to start back from Benton sometime in 1904. The local constabulary are taking a casual interest from the opposite platform. (Author's Collection)

the rest of the fleet had continued apace but deliveries of stock to Tyneside did not begin in earnest until the start of 1904.

The remaining fifteen vehicles built for the initial batch of NER electric stock were in small batches of specialist vehicles. All were driving vehicles and were 'double ended', meaning they had a driving cab at each end of the vehicle, like a modern mainline diesel or electric locomotive. They differed mainly in the type of seating they offered and in the case of the Diagram 100 Motor Luggage Vans this meant no seating as they had two empty spaces, a larger one for parcels and a smaller one for fish traffic. These, and all subsequent motor parcels vans built for the system differed in having two motor bogies, as they needed the extra horsepower to haul trailing loads up to 100 tons. One other difference these final fifteen vehicles had was the loss of the iron gates for passenger access from the platform to the vestibule and wooded half glazed doors fitted instead; the Diagram 90, 91 and 92 cars would be modified to match this change early in their lives. This was to be one of many modifications carried out on the fleet which will be explained in more detail later.

These were the initial ninety vehicles delivered for the start of full electric service in 1904. It became apparent from early on that further vehicles were going to be needed to cope with the increased passenger numbers, as patronage was quickly won back by the NER from the tramways. Between 1905 and 1915 a further thirty-five cars were constructed at York by the NER. They are detailed in the table.

A very rare photograph from the early years of operation. A unit is pictured at Central station but what is really of interest is the mesh screen over the driver's front window. This modification lasted a very short while and photographs of it are very rare indeed. (Author's Collection)

Table 2: Second batch of 35 electric cars built for the NER.

Diagram No.	Vehicle Type	Number/ Year Built	Car Number Range	Notes
106	Trailer Third	12 in 1905	3507–3518	
172	Motor Luggage Van	1 in 1908	3525	Larger fish compartment than Diag 100 car
175	Driving Motor Third	6 in 1909	3519–3524	Double driving end, Transverse reversible seats only.
192	Driving Motor Luggage Composite	2 in 1912	3770–3771	Double driving end, smaller than previous luggage compartment
203	Driving Motor Third	2 in 1915	3792–3793	Single driving end.
206	Trailer Third	11 in 1915	3781–3791	Reversible transverse seating only.
208	Driving Motor Luggage Composite	1 in 1915	3794	Based upon Diagram 192 but 1st class smoking compartment at end of saloon.

Commencing with No. 3519 in 1909 a change was made to the traction motors fitted from new, those being the slightly more powerful G.E. Type 211 rated at 150hp. The fleet reached its full complement of 125 cars with the completion of the single Diagram 208 vehicle in 1915, but from virtually the start of service the original fleet had been subject to modifications and these are certainly worth setting out.

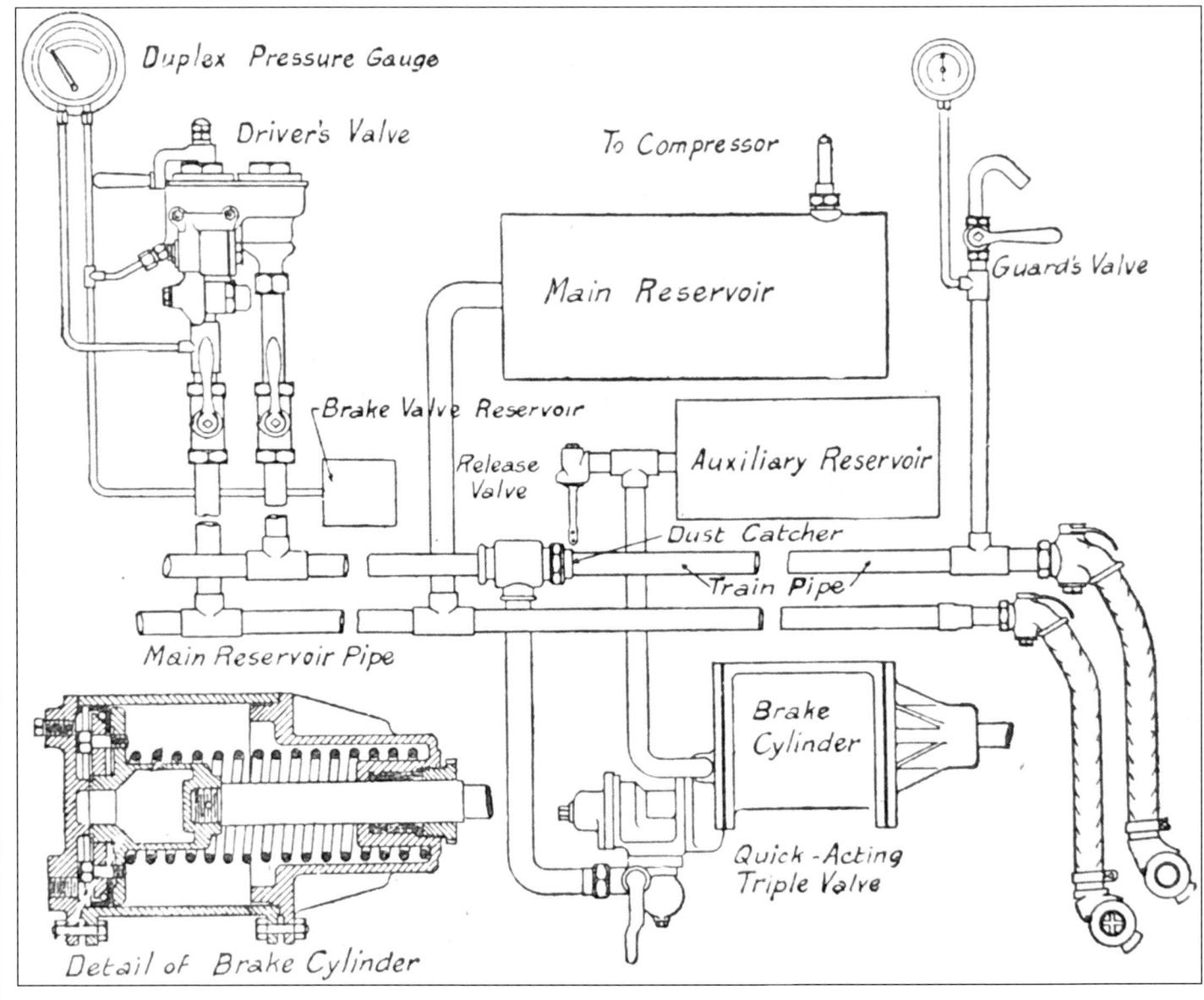

Diagram of the components of the Westinghouse automatic air brake as fitted to every vehicle of the NER electric fleet. (Author's Collection)

As mentioned previously, the iron gates that passengers used to access the cars from the platform on the initial three-car sets were changed to glazed wooden doors during 1905/6. At this time there was a serious rethink of the provision of first-class accommodation on the trains, with it being removed from all but the coastal express services from December 1904. Strong complaints led to this decision being reversed from April the following year but the NER decided to spread the first-class accommodation more evenly along the formation rather than concentrate it in one of the driving cars. Thus twenty-nine vehicles were converted to Driving Motor Composites with a mix of both first- and third-class accommodation. These were taken from the twenty-four Diagram 90 Driving Motor Luggage First cars which became Diagram 90 (mod) and the five Diagram 98 Driving Motor Luggage Third cars which came Diagram 98 (mod); additionally four of the

The First North Eastern Electric Stock • 33

The largest railway crossing in the world was the claim. This is the famous view of the Castle Keep in the V of the junction of lines to the north of Central Station. The electric unit on the left has been modified with the porthole driver's window and looks a world apart from the suburban steam service arriving on the right of it. (Newcastle Central Libraries)

Motor Luggage Van No. 3525 has the modified front window and is pictured at Walkergate at the head of the 'Controlled Set'. This rake of ten six-wheeled coaches was the second version that had been fitted with through wiring to enable the driver of the front van to control the rear van by remote control. The rear van is just visible on the left of the picture. (British Rail)

Diagram 90 vehicles were fitted with a second driving cab at the former non-driving end and these became Diagram 90A. This work was carried out between 1905 and 1908 and resulted in a subsequent modification to the 90A car, with the provision of a separate compartment for first-class smoking passengers. This was externally noticeable by the hinged top light windows to aid ventilation in this compartment.

In order to obtain more operational flexibility from the fleet, fourteen of the Diagram 92 and all twelve of the Diagram 106 Trailer Thirds were converted to Driving Trailer Thirds with

Luggage van 3267 survived in service as a de-icer van until the mid 1960s, numbered DE 900730. It was eventually saved for the nation but spent years in open storage at Walkergate waiting for a decision on any future restoration. This is where it was photographed in 1976. (Graham Pearson)

Above: **Luggage van** 3267 made it to the embryonic Stephenson Railway Museum in 1982, who had the unenviable task of restoring the vehicle after years of open storage. It is photographed here in 1982 shortly after delivery to Middle Engine Lane and is being assessed by the volunteers there. (Colin Alexander)

Right: **The best** things are worth waiting for. The fully restored 3267 was displayed outside in the sunshine when photographed here but is normally kept safely indoors at the Stephenson Railway Museum. (Brendan Gash)

the provision of a set of driving controls at one of the vestibule ends. The work was commenced in 1905 and required the provision of a cupboard on the end platform containing the power and brake controllers. The motorman would have a key to unlock this cupboard and access the controls when the vehicle was leading a formation. Lamps and destination indicators were provided on the roof end as per the other driving cabs, but there were no side windows. The Diagram 92 trailers thus modified were re-designated as Diagram 92a. Two further modifications followed to all of the driving cabs on the fleet. The first was the fitting of roller destination blinds in a box on the cab roof. This replaced the primitive glass plates displayed in the cab windows. The boxes also incorporated the lamps used for indicating route which were previously a separate stand-alone fitting. The second modification was the replacement of the full window at the cab front driving position with a wooden panel incorporating a circular porthole. At the same time as this was done the central doors on the front of the cab had their hinges moved to the opposite side away from the driving position. This may have been to prevent drafts and rainwater entering the cab area but no firm details were ever provided. Lamp brackets were also fitted to the sides of the cab fronts so the vehicles could carry standard oil-lit tail and head lamps should these ever be needed.

The whole fleet had to have improvements made to the underframes and window mountings from 1905 due to water ingress to the latter and sagging of the former. The frames had to be strengthened and the window mouldings changed to an increased depth of 1 1/2 inches. Other odd changes were made to several vehicles including one where the air compressor was moved from the first-class driving car to the third-class driving end; this did

The First North Eastern Electric Stock • 35

Above left: **The shoe** beams of 3267 still carry both the current collector shoes and the de-icing delivery shoes from its latter use. The quality of the restoration and the application of the original NER livery can be appreciated in this view. (Brendan Gash)

Above right: **Colour photography** didn't exist in the early days of NER electrics so black and white photos were retouched by post card manufacturers. This example is from 1904 and was produced by Bealls of Newcastle. (Author's Collection)

not cure problems with both noise and vibration so it was relocated back to its original position and changes were made to the way the compressor was mounted on the underframe instead.

Experiments were conducted on a couple of vehicles, neither of which were continued with beyond the initial fitment. One such was the trial of an electric door-opening arrangement on car 3243 which was tried in 1905 but was not considered to be better than the current passenger doors as fitted and was thus removed and never heard of again. Trailer car 3201 had small fans fitted inside the clerestory roof space to increase air flow for ventilation, something that was said to always blight the original stock. The fans failed to sort the problem and the fleet soldiered on with just the regular torpedo vents on the roof, although a few cars were fitted with opening top lights to alleviate the problem.

The fleet had reached its maximum size by the end of 1915, but events were to transpire to make that figure short lived.

This is my favourite of the retouched postcards that exist of NER electrics. Not only is the colouring to a very high standard but the photograph is from the earliest days and shows many features that would be changed on the units, such as the full size driver's window, the iron gates to the passenger saloon and the original position of the collector shoes on the end of the shoebeams. (Author's Collection)

Chapter 4
THE EARLY YEARS OF THE ELECTRIC SERVICES

As mentioned the first services were launched between New Bridge Street and Benton on 29 March 1904. This was intended as just a token launch and it has been suggested that the NER pushed to open this short section in an attempt to beat the Lancashire and Yorkshire (L&Y) company who were working on their own electrification project north from Liverpool Exchange. In the event the L&Y opened their first section a week before the NER but there is no hard evidence to confirm that there ever was a competition between the two companies.

The NER went for a phased introduction of their electric services after the initial opening. The first extension was from Benton to Monkseaton, which went live on 6 June 1904. The first through service left New Bridge Street at 6.50am and the rest of the day saw a 30 minute interval service with a service terminating at Benton in between those going the full distance to Monkseaton. At the time of the inauguration of this extension the NER was conceding that the work to switch on the full service was taking longer than anticipated but they remained optimistic that the opening of the next sections would now follow in quick succession. True to their word the service was extended to Tynemouth on 21 June, and on 1 July came the opening of the full loop into Newcastle Central and the Riverside branch via Byker. The first service left Central station at 4:35am with no ceremony and the new timetable became effective from then. Trains left Central for the circuit round to New Bridge Street at 05, 20, 35 and 50 minutes past each hour, with an hourly service for Tynemouth via the Riverside branch at 25 minutes past each hour. The Newcastle Chronicle reported that on the opening day most of the trains ran somewhat late but 'this was to be expected for such a new venture when the timetable is somewhat of an experiment.' The next part of the puzzle was added on 25 July 1904 when the line from Heaton to Benton East Junction was switched on. This section had no stations of its own but offered an alternative route for the circular services leaving Central for the coast, and on this date additional express services to the coast were introduced to the electric timetable. Not a part of the passenger story but very much a part of the story of NER electrification was the Quayside branch, a goods only line that used both third rail and overhead lines and was switched on in June 1905; more of this in a later chapter.

The final sections of the project were those that allowed trains to run in a loop out and back from Central station instead of New Bridge Street. The first modification was the electrification from Manors to New Bridge Street in 1909; this enabled the latter station to be closed and Manors North took over the role of terminus. The situation remained that way until March 1917 when the section between Manors North and Newcastle Central station was opened to passenger traffic, and services would only start and terminate at Newcastle Central's suburban platforms from here on in.

Much can be made of the impact the new electric service had on both the nature of commuter travel on Tyneside and the changes to operating practices in the area. But it must also be considered how huge the changes were for the staff who worked on the trains themselves. The transition from steam to electric was a great social change for the crews. It affected the way they dressed, the nature of the work they undertook and the health of those who had been raised on the soot and smoke atmosphere of the steam service. The electric trains required only one person in the cab: the 'motorman' (or driver as they would be called now) had to be retrained from the role of engineman and was now deprived of the company of a fireman as that role was redundant on the electric railway. The motorman now dressed in the same uniform as the rest of the train crew, as the need for boiler suit, grease top cap and overalls was a thing of the past. Now he wore a double breasted suit with a peaked cap bearing the company name; the motorman's uniform differed from that of the guard and the assistant guards only in bearing the legend of their respective job roles. The motorman had no need of shovels, oil cans or rags any more. The tools of his trade were the spanners required to unlock and operate the power and brake controls on the electric trains. He was issued with his own personal set of these and was responsible for keeping them safe and taking them with him when reversing a train that required him to change cabs. It is worth noting that all train crew were male. The advent of the female driver would not happen during the lifetime of third rail operation on Tyneside and female guards were still a number of years away too. The duties of the guards and assistant guards on the initial services involved the opening and closing of the iron gates that passengers used to board and alight from the train. They were also required to call out the name of each station when arriving at them; this was before any form of public address system of course. They also had to close the gates when ready for departure, and whilst the train was in motion they could only ride in the third-class cars to shelter from the elements.

It was the welfare of the motorman that was behind the earliest and most noticeable modifications to the original fleet. During the days of steam the crew on the footplate of a locomotive were protected from objects, such as stones, being thrown at them, by the sheer bulk of the boiler or coal bunker, depending on which way the locomotive was facing. But on the electric units there was only a glass window between them and the missiles and when hit the window could smash, with the prospect of shards of glass injuring the motorman. According to sources it was the trade union that requested that the front end offer better protection against stone throwing. The NER at first affixed a steel mesh over the driving cab window but this was not an ideal solution, and so to reduce the amount of glass in front of the motorman and thus the size of the target the driving cab window was changed from a large rectangular one to a round porthole. This change was effected during 1906–1908 on all units and gave the NER stock a distinctive front end that was different to other electric units of the time. The result of the modification was that the motorman usually had to stand when driving to be able to see out of the front window.

Consideration was given to the fitting of lavatories to the fleet very early on and the NER's Carriage & Wagon draftsman, Mr Groves, came up with a set of plans for this that involved options for either one or two lavatories on each three-car set. The proposals passed across the desk of Vincent Raven but were never proceeded with. It was not the custom of the NER to provide such facilities on suburban routes.

Passengers are ready to board the service to Manors North as it enters Backworth station. The picture is dated c.1915 and by this time New Bridge Street had closed to passengers and Manors was the destination of anti-clockwise coastal circle services. (Author's Collection)

Given the variety of vehicles in operation on the system there was no set formation for any services. Whilst the three-car set was the basis and the sets were marshalled and numbered from 1 to 25, additional single vehicles could be added and trains could be run formed of anything from two to six cars in length. One oddity that began to appear on the network from 1905 was the controlled set. This started life as a formation of six steam stock trailer coaches that were hauled by two of the Motor Luggage Vans. The necessity for this had arisen due to the ongoing modifications to the fleet that left a shortfall in the number of available sets for traffic. This temporary measure became more permanent in 1908 when the six trailers were strengthened to become a formation of ten six-wheel carriages of Victorian vintage, with through cables provided so that one of the Motor Luggage Vans could be marshalled at each end and both were under the control of a single driver. This oddity remained in use on workmen's traffic, principally during peak hours on the Riverside branch up until 1929, when a second controlled set was created to replace the earlier one and was formed of six bogie carriages.

Another aspect of how the introduction of the electric service changed the railway significantly was most evident at Newcastle Central itself. Because it was both a hub of east coast express traffic and a terminus for local commuter, coastal and secondary routes, the station was never quiet and the number of train and locomotive movements witnessed hourly at the turn of the century was simply staggering. Whilst through traffic simply arrived and departed with the usual bustle of passengers and parcels being loaded and unloaded, along with any locomotive changes that were required, it was the local traffic terminating and restarting

Something appears to have gone wrong with this electric unit just outside Central Station. With the Castle Keep on the right and plenty of personnel in evidence on the ground it is not clear what has caused this service to come to a halt. There are plenty of bemused passengers on board too. (Newcastle Central Libraries)

from the station that clogged the track layout. For every train that arrived, a second locomotive had to be brought into the station to haul the return service away, and then the original locomotive that was then freed from the buffer stops had to exit the station to be stood off to take on water and coal and await its next working.

The electric services introduced a new type of train to the area, one we now call the multiple unit, so called because any number of vehicles can be coupled both physically and electrically and all the motors on the train are under the control of the single driver in the leading cab. The basis of the system was designed in America by Frank Sprague who was working on a system to control multiple lifts in skyscrapers. He discovered his developments would have equally far reaching effects in the emerging the world of electric railways. Now there was no need for a separate locomotive as all the traction equipment required for motion was carried on the carriages themselves, and as soon as a train had entered the terminus all that needed to happen was for the motorman to walk back to the opposite end, remembering to take with him the spanners for operating the power and brake controllers. After setting up the cab the train was ready to depart on its next working. No second locomotive was required, no coaling stage, no water tower and only a single in and out movement instead of several that had been required under the days of steam. Whilst Newcastle Central remained an extremely busy station, the pressure on the track layout was considerably reduced with electrification, and this played a big part in how the layout would be rebuilt over the following decades.

In 1904 electricity was not widely understood by the population. One has to remember that the streets were still lit mostly by gas and horses provided most of the power for road transport. The emerging science was not widely taught in schools. Despite the cast iron warning notices the NER placed at intervals around the network warning of the dangers of straying onto the electric lines, it was only a matter of time before the live conductor rails would claim their first victim. On Wednesday 30 August Joanas Whitehead, a 12-year-old boy from Byker was heading to Dent's Hole to catch crabs in the river with friends; they had decided to take the short cut on the way back and scaled a wall near St Peters station to get onto the railway line. It was whilst crossing the line that young Jonas tripped and fell on the running and conductor rail, thus receiving the shock of the full 600v DC. He was killed outright, and his body was removed from the track by the crew of a steam engine that was working the line. This happened on the Riverside branch which was not open to electric trains at that time but was under test prior to full commissioning less than a month later. Two days later a couple of schoolboys had a lucky escape when they were found to be throwing their caps onto the live rail at Walker then climbing a wall onto the track to collect them. On this occasion the power was switched off but the boys were apprehended and fined 2s 6d each at the Newcastle Police Court. It was also recommended by the court that they read the report into the death of Jonas Whitehead so as to teach them the error of their ways.

By 1913 the NER had recorded over 10 million passenger journeys for the year on the electrified network. This marked a complete reversal of the huge decline that had prompted the project in the first place. They had taken on the trams and they had won. The fleet reached its full size of 125 vehicles in 1915 and by March 1917 the third rail was laid and operational over the full area the company had intended, after deciding to abandon plans to electrify the Ponteland branch.

On 11 August 1918 the system endured an infamous day that was to have ramifications for the next couple of years. At around 07:00 on that Sunday morning a fire broke out inside the shed at Heaton, where over half of the fleet of electric stock was berthed. It being a Sunday there was less requirement for stock to be in service and this was a contributing factor to the extent of the ensuing damage. The origin of the fire was never properly established but what is known was that it took hold very quickly and staff on duty had to act to get as many trains out of the shed as they could. It was estimated that over fifty carriages were saved from the flames by the actions of staff, and one Foreman, James Bailey, had to be taken to the Newcastle Infirmary with severe burns. The NER's own fire brigade attended and were shortly joined by Westgate and East End fire brigades. Between them over 1,000 yards of hoses were set up to tackle the blaze using whatever local water supplies

A two-car set on the Manors North to Benton shuttle service pictured at West Jesmond station. (Author's Collection)

Another view of the famous crossings outside Central. But there has been a change of livery for the electric stock since the last such image. The NER units on the right now wear the all over crimson lake livery, meaning this image is post First World War. (Newcastle Central Libraries)

they could connect to, as the supply at the shed itself was described as totally inadequate. Reports describe the shed as being alight from end to end and it was believed the stores of oils and cleaning materials used for servicing the stock acted as an accelerant which aided the fire to take hold of the structure so quickly. When the fire had finally been extinguished the shed was completely destroyed, and so too were thirty-four carriages of various types. The NER estimated the cost of the damage at between £150,000–£200,000. The next day seven sets of steam hauled carriages, together with locomotives and crews, had to be drafted in to provide a full service. It would take a month for the NER to regroup and provide a full electric service again, but with the compromise that several services had to be formed of shorter trains than previously. The full list of vehicles destroyed in the fire is set out in the table.

Table 3: Vehicles destroyed in the 1918 Heaton shed fire.

Diagram Number	Vehicle Type	Number Destroyed	Destroyed Car Nos.
90	Driving Motor Luggage First	5	3243/6/9/50/1
90a	Driving Motor Luggage Composite	1	3241
91	Driving Motor Third	7	3217/9/21/5/6/7/8
92	Trailer Third	1	3180
92a	Driving Trailer Third	3	3195/97/98
97	Driving Motor Third	1	3268
98	Driving Motor Luggage Composite	1	3238
99	Driving Trailer Third	1	3211
106	Driving Trailer Third	5	3507/8/12/3/8
175	Driving Motor Third	3	3520/21/24
202	Trailer Third	1	3204
203	Driving Motor Third	1	3212
206	Trailer Third	4	3785/8/90/91

In addition to those vehicles completely destroyed there were a number that were damaged but repairable, and others suffered smoke damage but could be repaired. The NER now had to consider how to move forward from this setback. The Tyneside electric services had been a huge success and the NER were not going to let that get away from them.

Chapter 5
THE FIRE REPLACEMENT STOCK

The NER held meetings to discuss how to manage the aftermath and impact of the Heaton car shed fire. Besides the extensive damage to the shed and facilities, the rolling stock losses had been considerable. A total of thirty-four passenger vehicles, which accounted for over a quarter of the fleet, had been either destroyed or were beyond any form of repair. Others had been damaged to a lesser extent; some suffering only smoke damage, broken windows or exterior tarnishing, and these were hastily repaired and put back into traffic. The carriage shed at Heaton was also destroyed beyond economic repair and a decision had to be taken regarding the replacement of that facility, either on the same site or elsewhere. In the end the NER opted for the latter with a new carriage shed for the electric fleet planned on the north side of the network at South Gosforth. Work on its construction was authorised in August 1920 but would not be completed until after the grouping of 1923. In the meantime the Heaton site was converted to stabling sidings with a simple three road shed provided for maintenance of the fleet in the interim. The operation of the electric network would be strained for the next five years as it had been robbed of not only over 25 per cent of its major resource, namely the trains themselves, but also the dedicated facility that looked after the surviving vehicles. What had not been lost in the fire was the system's staff. The workforce of the NER was notably proud of their network and given that the fire happened during the last few months of the First World War there was a heightened sense of duty across the country. It would be this resolve and spirit that would be a contributing factor in keeping the service running during these difficult times. It may also have contributed to an understanding amongst the NER's customers that the company was doing its best despite the adversity.

There was no doubt that new cars would have to be built to replace those lost in the fire, and construction of what was to become known as 'The Fire Replacement Stock' was authorised very quickly. The work of construction fell once more to the NER's facility at York. The replacement stock was ordered on an almost like-for-like basis to the vehicles that had been destroyed, so nineteen driving motor cars and fifteen trailers were ordered; the difference was that the trailers were not built with driving positions despite that being a feature of several of the cars lost in the 1918 fire. This may have been to speed up the design and construction work, as the fewer variations to the build requirements could save valuable weeks and even months in the process to deliver completed cars. The new vehicles were even given the running numbers of the burned-out vehicles they were replacing. One additional vehicle was added to the order: a single double ended Motor Luggage Van, despite none of the original batch having been lost in the fire. It may have been the case that these vehicles had proved so useful in the network's operations that the chance to boost the fleet could not be overlooked. The summary of the replacement vehicles built is set out in the table.

Table 4: NER Fire Replacement stock.

Diagram No.	Vehicle Type	Number Built	Running Numbers
218	Trailer Third	15	3180/95/7/8 3204/11 3507/8/12/3/8 3785/8/90/1
219	Driving Motor Luggage Composite Double Ended	2	3238/41
220	Driving Motor Luggage Composite Single Ended	5	3243/6/9/50/1
221	Driving Motor Third Single Ended	5	3212/7/9/21/5
222	Motor Luggage Van	1	3795
223	Driving Motor Third Double Ended	7	3226/7/8/68 3520/1/4

Appearance wise the replacement stock followed the 1904 stock in many respects. The same underframe and body dimensions were used and the vertical matchboarding of the lower body side remained a feature as did the timber sheeting for the remaining body sides. Lessons learned from over fifteen years of operation of the original units were incorporated into the replacement stock.

The most noticeable difference was the loss of the clerestory roof in favour of a semi-elliptical one. This feature gave the outline of the cars a more 'British' appearance and moved ever so slightly away from the American look of the original cars. The poor ventilation of those original cars meant the new batch was fitted with hinged top lights above the main side light windows. These could be opened or closed by passengers as required to improve the air flow inside the car. Two rows of torpedo ventilators were mounted on the car roofs, one either side of the centreline. Reversible rattan covered seats in the same style as the original stock were retained as were the leather covered first-class ones, while heating and lighting as before were electrical. One improvement on the previous builds was the fitting of larger motors, these being supplied by British Thompson-Houston (BTH) and were their Type 254B motor rated at 140hp. The original stock had never proved underpowered, but traction motor design was an ever improving area of technology and the NER was simply using the best that was on offer at the time. As the replacement stock had to work in mixed formations with the older cars the new batch was fully electrically compatible and had the same electrical connections and Cowhead couplers as had been used before. The cab ends carried both the roof mounted roller destination blinds and the integral marker lamps in the headcode box as per the previous stock. The brass whistle was roof mounted too. The vehicles were delivered between 1920 and 1922 and were painted in an all over crimson lake livery with the company name, numbering and lettering in gold at mid level on the matchboard panels. The NER crest appeared either side of the company name and in due course this colour scheme would be adopted and applied to the earlier-built vehicles of the Tyneside electric fleet.

The single Motor Luggage Van was built in 1922. Internally it was based on the Diagram 172 vehicle with a larger 28ft luggage compartment and smaller 22ft fish compartment. It also had conventional side buffers so it could haul van traffic, and was equipped with two motor bogies. Apart from the semi-elliptical roof, one significant difference from the three earlier vans was the provision of a vacuum exhauster so it could haul vacuum-braked stock; this equipment was mounted slung from the underframe. This made sense as all of the NER's passenger rated luggage vans and an increasing number of goods vans were equipped with only vacuum train brakes. The van was altered internally in 1938 to make both compartments of equal length.

The 1920, or Fire Replacement Stock, had been born out of both

A Driving Motor Luggage Third of the 1920 fire replacement stock in original crimson lake NER livery. (Author's Collection)

Third-class interior of a 1920 stock car; the seats are reversible. (Author's Collection)

necessity and adversity. The NER had proved it was resourceful in building these units so quickly when there had been no forward plan to do so and resources had to be used all the way from draughtsman to procurement and workshop craftsmen without any adverse impact on the NER's existing rolling stock and motive power construction and overhaul plan. It had also proved the NER's openness to new ideas and how the replacement stock had been viewed as an opportunity to improve upon the vehicles they were replacing when it would have been quicker to simply dust off the plans for the original cars and simply say 'we will do that again'. In the event the 1920 stock would go on to be some of the longest lasting in the history of the Tyneside electric network.

The completion and delivery of the replacement cars would bring the system back up to the full strength as envisaged by the NER prior to the First World War. The replacement cars were simply integrated into the service pattern and mixed rakes of original and replacement cars were quite the norm. As the cars were fully compatible and near identical replacements for the lost vehicles, the service was finally able to operate as the NER had intended it to prior to the fire. It had been a taxing and gruelling struggle to deliver that vision, but the NER had persevered against adversity and got to the finish line. It was ultimately to be their swan song, for the days of the NER were numbered. All of Britain's railways were still feeling the effects of those tough war years which had placed great strain upon the infrastructure of all networks, with the need to be a prime mover for material, men and commodities vital to the war effort. The economic burden placed upon what were private enterprises that had never been set up to cope with such demands had been enormous and a major change took place at the start of 1923. By an Act of Parliament all the railway companies were condensed into the 'Big Four' and the responsibility for the Tyneside Electrics passed to the London and North Eastern Railway, who would carry on with the operation and development of the fleet for the next twenty-five years.

Chapter 6
THE LNER TAKES OVER

Monday 1 January 1923 was the start of a new era for Britain's railway networks, which had been used and abused during the four years of the First World War. A lack of investment in maintaining the infrastructure, motive power and rolling stock due to the demands of the nation's war effort had left the system 'running on fumes'. It was painfully apparent at governmental level as much as it was to the officers of the dozens of railway companies themselves that a massive change had to take place for the railway network to recover and provide what was required of them as the country returned to peace-time rebuilding and the much hoped for era of prosperity. The government fell short of a full-blown nationalisation of all the railway companies but instead opted for a halfway house solution where operators were consolidated into four large companies based upon regional operations. These companies remained private businesses run outside of government and absorbed the assets and responsibilities of their individual constituent concerns. In the case of the North Eastern Railway this meant they became a part of the London and North Eastern Railway (LNER).

The NER was one of the larger and more significant portions that made up this new concern. The NER had worked closely on through services with the other railway companies that were thrown into the new business alongside them, such as the Great Northern Railway who operated from London to York and the North British Railway who handled through traffic from the NER into Scotland. But from now on there was only one company, and whilst it would take many years for individual loyalties to accept this internally, outwardly there was a new corporate identity and common purpose.

The LNER's flagship services were the East Coast express workings from Kings Cross to Edinburgh Waverley. It was these that had the fastest, most impressive locomotives and the best carriages, and were the focus of so much of the company's marketing, the most famous of which was the Flying Scotsman service. Away from all of this main line glamour the LNER was not the most lucrative of the big four railway companies and to turn a profit it would be reliant on both heavy freight flows, notably the coal traffic from the north east and Yorkshire coal fields, along with the commuter traffic around both London and Newcastle, together with a few of the other major cities within the LNER network such as Leeds and Edinburgh. These were important revenue streams that were not seasonal but provided an all year round cash turnover, and the Tyneside electric network was the only one so far electrified, as neither the North British Railway nor the Great Northern Railway had been inclined to commit to any such works on their suburban routes.

One of the first acts of the LNER was to open the new carriage shed at South Gosforth that had been commissioned under the NER to replace the fire damaged shed at Heaton. It was opened on 1 October 1923. The main carriage shed comprised a brick built structure with ten roads, each capable of holding twelve cars. The building was purposed for berthing, cleaning and maintenance of the electric fleet, with an additional two road repair shed being built on the north side of the running shed that had an overhead crane and inspection pit so heavy repairs could be undertaken without the regular requirement to send vehicles away to York works. The running shed was double ended, so trains could enter or leave from either end of the building, and situated on the Gosforth Loop between Benton and West Gosforth stations on the

A mixed formation of stock at Heaton. The leading vehicle is 1920 fire replacement stock, the second is a clerestory roofed car, and the remainder of the train is 1920 stock. (Author's Collection)

A North Eastern Railway EMU set on a South Tyneside service to Newcastle Central in c.1930 (location unknown). (Rail Archive Stephenson)

north side of a triangle of junctions and not enclosed by them, so operationally it was a very flexible location.

Under the LNER the core service on the Tyneside electric routes remained unchanged from that which had been provided under the NER. The most visible change was the repainting of all the cars from the NER's crimson lake livery to a rather drab teak livery. All the lettering and numbers on the vehicles continued to be in gold. The LNER introduced a new numbering policy in April 1925 which meant a 2 was added to the start of all the car numbers of the electric stock. Prior to this the NER car number under the LNER had a simple Y suffix to denote it was a York area vehicle. The new system meant the letter suffix could be dispensed with.

Accidents and collisions occurred occasionally on the system and these were of varying degrees of severity. Some were collisions with the buffer stops at the terminals at New Bridge Street and Newcastle Central stations. Other accidents involved collisions with either trains or locomotives. Some resulted in damage to the stock that could be repaired and some involved injury to passengers and staff. The first fatal collision occurred in the early hours of 18 December 1923 when an empty train leaving South Gosforth shed collided with a steam engine over a road bridge. The force of the impact knocked the engine off the bridge and its tender landed on top of the cab, killing both the driver and fireman.

The most bizarre and tragic accident on the system during the LNER years occurred on 7 August 1926, when a passenger train collided with a goods train at Manors. What made this accident so alarming is that the electric train was under power but had no driver at the controls at the time of the collision. The official inquiry revealed that the driver had tied a piece of rag over the dead man's handle, a safety device that normally had to be held down with natural forearm pressure whilst the train was in motion. If it was released the automatic brakes would apply. By tying the push button down and rendering this safety feature inoperative the driver had been able to leave the cab of his still moving train and had gone back into the luggage compartment behind to lean out of the exterior door and observe passengers in the saloon area behind. Quite what his motivation for this was we shall never know for at some point near Heaton he struck a bridge which pulled him from the train and killed him. The undamaged train carried on out of control, with the safety device tied down and inoperative. All that was left to stop the train was the impact with the goods train that has been signalled to cross in front of it at Manors. Remarkably no passengers or staff were killed, other than the foolish driver. Three of the leading cars of the train were badly damaged, two of which were repaired in due course; but the leading car, 23253, was withdrawn and a replacement ordered from York which was constructed and delivered in January 19. This was a Diagram 220 vehicle and was allocated the number of the withdrawn car.

The controlled set of antique six-wheeled hauled coaches that operated with a Motor Parcels Van at either end was replaced by the LNER in 1929 with a new set of six-bogie carriages. As with the original set the coaches were converted to be sandwiched between two parcels vans and the necessary through wiring was provided to enable one driver to control all the motors on the train. Five of the coaches were eight compartment third-class coaches with the sixth being a composite coach. The set was utilised mainly on Monday–Friday traffic for workmen on the coast and Riverside loop. It was pressed into action on Saturdays on a single service from Walker Gate to Monkseaton via Tynemouth and would be available for extra traffic as seasonal demands on the routes dictated.

An original NER cast iron electrification warning sign on the lineside near Whitley Bay. (Graham Pearson)

The 1930s were to be a testing decade for not just the LNER or the people of Tyneside but also the world as a whole. In 1929 a global financial crisis occurred when the stock market on Wall Street crashed, wiping millions of dollars from the value of shares. Trading in stocks and shares had become a popular source of investment and many people had virtually treated it like a hobby, borrowing heavily to play the markets. With the value of their shares now a fraction of the value of their loans thousands of people became bankrupt literally overnight. The economies in the both the United States and Europe crashed and the effects began to filter down to the working classes who found themselves out of work in increasing numbers. This period is nowadays referred to as The Great Depression.

On Tyneside the effects of the depression began to be felt: several of the area's ship yards were closed as there was no longer the same level of demand for ships to be built, while coal continued to be mined but demand fell as industries tightened their belts and relied on their stockpiles before buying more materials in. As one major employer closed, the knock-on in the area would be felt with dozens of smaller businesses also succumbing as they relied on the men and women who spent their wages coming through the doors to survive. Those small concerns now found their customers no longer had a disposable income, or any income at all for that matter. The LNER were faced with a dilemma. The system was still providing a well used service but there was little revenue being created that would allow for growth and upgrading of the infrastructure, something that was now a serious consideration as the network was approaching thirty years of continuous operation. If the depression was a short term problem that would eventually be overcome, then the LNER had to be ready to meet the return of an increasing demand that would inevitably follow. Indeed, new homes were being built in the area and this was likely to bring fresh patronage. Two projects that did get the go ahead were the opening of a new station at West Monkseaton, which was opened on 20 March 1933, and the alterations to the lining of the tunnel at North Shields where trains were not allowed to pass due to tight clearance. The tunnel walls were trimmed, the track bed lowered and the lines slewed appropriately so the restriction was lifted. Recognising the need to safeguard its precious commuter traffic, in June 1934 the LNER agreed to make a significant investment and take the risk of replacing the clerestory roof stock, announcing an order for forty-four two-car units of an articulated design along with two single luggage vans. The 1920 stock would be refurbished to work alongside the new trains. In 1935 government policy allowed this plan to be considerably expanded.

Chapter 7
OLD TRAINS, NEW ROUTE

In an effort to stimulate the economy during the severe economic downturn of the early 1930s the government of the day had decided to make available loans to large industries at highly competitive rates of interest, provided the funds were used for new construction and improvement projects. Government had recognised that the lack of faith in the economy following The Great Depression had meant a reluctance for private finance to be made available, and in many cases it had become harder to secure. This impasse was stunting growth and with it a return from the period of economic recession.

Several railway companies took advantage of the new scheme to improve their routes with projects that encompassed both electrification and new rolling stock. One such project that had lain on the shelf for a year was the 1934 proposals for the Tyneside electric network which the LNER re-visited and expanded way beyond the scope and ambition of their initial report. With finance from a Government loan the plan now included in the first part the replacement of all the earliest electric stock on the Tyneside network with a fleet of sixty-four articulated two-car units and four single vehicles. The 1920-built 'fire replacement' units were to be refurbished and put to work on a newly electrified route from Newcastle to South Shields, south of the River Tyne.

The railway first reached South Shields in 1834 as part of the Stanhope and Tyne Railway. Extensions and linking up with other routes meant the town had a direct rail connection to Newcastle by 1848. The following year Queen Victoria performed the formal opening of the High Level Bridge over the River Tyne linking Newcastle and Gateshead. This structure had two decks, a lower one for road traffic with the rail deck above it. The completion of this bridge meant the line as it would eventually become electrified was in place. There were additional stations opened on the route over the next three decades and its importance as a source of freight and passenger traffic had grown incrementally during those years. By the 1930s the route had a strong mix of residential areas and industrial sites that could provide commuter and workmen's traffic as well as the potential for off-peak travel to South Shields for those wishing to spend a few hours away from the bustle of the city.

This was not the first time that electrification south of the river had been considered. Several reports were commissioned by the NER in the years before the First World War looking into the feasibility of not only electrification to South Shields, but also to Sunderland, the line to which diverged at Pelaw Junction. The reports were favourable but the project never got acted upon and with the advent of the war, followed by the grouping of 1923, the idea had been put aside. The LNER had made improvements to the power supply for the Tyneside routes by installing new equipment in the substations at Cullercoats, Benton, Pandon, Wallsend and Percy Main. This took the form of Metro-Vic built rotary convertors to replace the original equipment installed for the NER. This renewal was supplemented by new converter units being installed at lineside huts at Earsdon Grange and Gosforth East. These two sub-stations were of single converter type and served to 'top up' the DC voltage which, in common with all conductor rail systems, was prone to earth leakage and subsequent voltage drop due to its ground based nature and exposure to the elements. This new equipment was barely broken in when it was replaced by the Central Electricity Board at no cost to the LNER due to a change in the AC cycle frequency introduced in the area, which meant the equipment installed by the LNER was no longer suitable for converting the AC current supplied by the Central Electricity Board to direct current for traction use.

The contract to construct the new units for the North Tyneside routes was awarded to the Metropolitan-Cammell Carriage & Wagon Company Ltd who would build them at their works in the midlands. This was, of course, a departure from previous stock which had been built at the railway company's own works at York. Tenders for the construction of the fleet had also been received from Gloucester Carriage & Wagon Co and from the Birmingham Railway Carriage & Wagon Co, both of which had bid too high and were passed up for the work. Met-Cam had considerable experience of building EMUs as they had supplied vast numbers to the London Underground and the LMS as well as several export orders over the previous two decades.

A two-car set of 1920 stock after refurbishment for use on the newly electrified South Shields route. (British Rail)

Driving Motor Composite 24465 after completion at York in 1938. This was the last Tyneside electric vehicle built at York. (British Rail)

Due to the need to form the 1920 stock into eighteen two-coach units for use on the South Shields line, York was given the chance to build one final car for the Tyneside electric routes. This was because thirty-five individual vehicles existed and an additional one was required to make up the numbers. Car 24465, a Driving Motor Composite built to Diagram 281, was rolled out of York workshops in 1938. Around the same time as this vehicle was being constructed the other vehicles of the 1920 stock visited York to undergo rebuilding for their new role as the South Tyneside electric stock. The fleet would eventually comprise eighteen Driving Motor Coaches and eighteen Driving Trailer Coaches. It was necessary to de-motor three Driving Motor Coaches, one single cabbed and two double cabbed, to increase the number of driving trailers to eighteen. The Driving Trailers, including the three former motor coaches, were grouped as Diagram 218 (modified) and had non-reversible bucket seats fitted upon refurbishment as well as driving controls in a lockable cupboard at the vestibule end (except the two cars originally built as double-ended driving motors). As fifteen of these vehicles were built as pure trailer cars these had to have a normal driving cab fitted at one end. When complete each vehicle had a seating capacity of sixty-eight third class. The three former Driving Motor Coaches differed in their capacity due to their original layouts: car 23217 had sixty-four seats whilst both cars 23268 and 23524 had sixty third-class seats.

With one new vehicle under construction this left seventeen Driving Motor Coaches built to five differing diagrams to be refurbished for services on South Tyneside. Seven of these were double-ended driving cars that had the driving controls removed from one end as it was decreed that all South Tyneside Driving Motor Cars would be single ended. Eight of the vehicles were first/third composites and were to remain as such in their new role, whilst the nine Driving Motor Thirds did not gain first-class accommodation. So after the new build vehicle was constructed this provided a final fleet comprising of nine Driving Motor Thirds and nine Driving Motor Composites. Bucket seats were fitted in the third-class seating saloons but leather upholstered seating was continued with in first-class saloons. The whole fleet was repainted into a livery comprising red lower body side and off-white upper body with lettering and numbers in gold, more or less the livery of the original NER. All underfloor equipment and running gear was finished in black. Due to the reduced number of first-class vehicles and the need to make these more easily identifiable to customers at platform level the cab ends had a small horizontal black stripe added half way up the white section of the body side between the cab front and side windows.

A train of 1920 stock at Newcastle Central with a service for South Shields in the early 1950s. The unit has been repainted in BR green livery.
(SERA Archive)

South Shields service formed of 1920 stock EMU enters Gateshead East station. (R. F. Payne/ARPT)

A snowy Newcastle Central station on 7 February 1954. The mechanical signalling and the impressive signal box are evident. (J. W. Armstrong/ARPT)

On the ground the work to electrify the line to South Shields had begun in 1935. The South Tyneside electric services were to use the same electrified suburban platforms at Newcastle Central. From there the route passed out of Central station, taking the right-hand fork of the diamond crossovers (to the right of the castle) and then over the High Level Bridge to cross the River Tyne onto its south bank at Gateshead station. Here the conductor rails were only installed at the two platforms on the east side of the station that served the South Shields line and not at the other two on the west side of the station which were for the Team Valley route to Chester-le-Street and Durham and were served by only a handful of services each day. From Gateshead the line headed due east through Felling and Pelaw station; the later was where the junction and passenger interchange for the line to Sunderland was located. The next stations were Hebburn and then Jarrow. Jarrow had come to symbolise the era of the Great Depression when during March 1936 200 men marched from Jarrow to London to protest at the unemployment and poverty that was being experienced in the area following the closure of the town's main employer, Palmer's Shipyard two years earlier. The 'Jarrow Crusaders' were met with indifference when they arrived in London to deliver their petition, but their actions have left a lasting legacy and the name of the town will always be associated with their actions.

The route then passed through Tyne Dock and High Shields stations before the terminus at

South Shields was reached. The station at South Shields boasted two curved platforms with a third track between them to facilitate running round locomotives so they could haul their train back to Newcastle. An ornate canopy covered the middle third of the platforms, with a lattice work iron footbridge enclosed within the roof to facilitate access to the two platforms. Despite being a passenger terminus the railway continued beyond the station to serve coal staithes. This was the fourth station to be opened at South Shields and dated from 1879; the previous three all had short lives and were replaced as the railway in the area expanded and was remodelled. As events would latter transpire this would become a theme of stations at South Shields. The same year that the latest South Shields station opened the preceding calling point of High Shields also got a new station: two platforms with NER brick buildings provided at street level and on the up platform (up trains went to Newcastle) where a glazed canopy was also provided. If anyone was in doubt about the up and coming importance of the South Shields line in 1879 then these new stations should have convinced them that this was a line with a future. Electrification would further enhance the importance and relevance of the route in much the same way as the station buildings had done nearly sixty years earlier. Additional electrified open air berthing sidings were provided at Pelaw and also at South Shields, beyond the station adjacent to the aforementioned coal staithes, where stock could be berthed both overnight and between the busy

South Shields service departing Newcastle Central. The horizontal line to the right of the cab window denotes that first class seating is in that end of the train. (Author's Collection)

This South Shields service is crossing the High Level Bridge and enters its first station stop at Gateshead East. (J. W. Armstrong/ARPT)

Newcastle Central station 29 June 1954, and this train of 1920 stock has less than a year left in service on the South Shields route. (F. W. Hampson/ARPT)

morning and evening peak periods when it was not required for the service.

The service would be worked by the two-coach units which could be marshalled as either two-, four- or even six-car trains as demand at various times dictated. The platforms at Hebburn, Jarrow, Tyne Dock and High Shields all had to be extended to cater for the longer electric services. Four substations were constructed to feed the new conductor rails at Gateshead, Pelaw, Jarrow and Tyne Dock. These installations were fitted with mercury arc rectifiers and rotary converters. The substations were under remote control from the new electrical control room north of the river at Wallsend which also oversaw the supply to the North Tyneside lines. Due to the change in AC cycle supplied to the LNER by the Central Electricity Board to 50 cycles per second the DC voltage of the conductor rail on both sides of the river was set at 630v DC.

The train service was to be run at twenty minute intervals with a Sunday service every half hour. The electric units replaced a service that previously required upwards of

Old Trains, New Route • 55

No. 29187 leads a train of BR green liveried 1920 stock on a South Shields to Newcastle Central service entering High Shields in September 1954. (Colour Rail)

thirty tank engines and two dozen carriages, all of antique vintage, that had previously been run at mostly twenty minute intervals, but not as a regular clock face timetable due to the need to change engines during the day for water and coal replenishment. All works were completed in March 1938 and an inaugural train ran out and back to South Shields from Newcastle Central on 14 March. The full electric service was introduced on the route on 2 May, and calling at all stations took thirty-three minutes from end to end.

Chapter 8
NEW TRAINS, OLD ROUTES

Whilst the electrification work of the South Shields Line and the refurbishment of the fire replacement stock that was to work it had been progressing, further south the construction of the new stock for services north of the river had been progressing. The LNER ordered 132 new vehicles; the general specification for the fleet had been issued by the Chief Mechanical and Electrical Engineer's office under the auspices of Nigel Gresley, and from this Met-Cam were given free hand to make the final design. Of the vehicles to be built only four of them were to be independent cars, the rest would be formed into sixty-four two-car articulated sets. These sets were divided into four different types that could be marshalled together to make up longer train sets as service demands dictated. Each set was identified by a letter, either A, B, C or D. There were a number of new features for the region introduced with this new build but some of the flavour of previous stock was perpetuated, with the retention of Cowhead couplers at the driving ends and hand-operated sliding doors to the passenger vestibules; the LNER opting not to go for the air operated doors that had become standard on EMUs of both the LMS and London Underground lines. All cars were mounted on welded steel underframes with bodies made of riveted steel and each one measured 55 feet and 9 inches in length. Three different types of bogie were found on each two-car set. The driving end of each motor coach had a powered bogie with an 8 feet 6 inch wheelbase with each axle fitted with a 216hp Crompton Parkinson traction motor. These motors gave the trains a maximum design speed of 55mph, sufficient for the start-stop nature of the work they were to undertake. The other end of each set had an unpowered trailer bogie with a wheelbase of 8 feet exactly. The third bogie was shared by the inner ends of each vehicle due to the articulated nature of their construction and had mountings to support each vehicle, so had to be of heavier construction than the outer trailer bogie. It did have the same 8 foot wheelbase though. All wheels, both motored and non-motored, were of 3 feet 7 inch diameter. Both the outer end bogies were fitted on each side with a wooden beam that supported a single collector shoe at its midpoint.

All sets were fitted with two braking systems. The standard Westinghouse twin pipe air brake was the automatic service brake giving the trains a fail-safe brake that would operate in the event of a train division or the operation of an emergency handle by either a passenger or guard.

In addition to this a non-automatic electro pneumatic (EP) brake was fitted that was to be the primary brake for smooth stopping at stations and signals. The EP brake worked by having an electrically operated valve that opened when the motorman moved the brake controller to the 'EP application' position, and thus allowed air from the train's reservoir system to enter the brake cylinders. When the motorman moved the brake controller back to the 'EP lap' position the air in the cylinders was held there and could be added to by repeating the process. The air in the brake cylinders would only be fully released when the motorman replaced the handle to the 'running and release' position. One of the advantages of this system was that due to its electrical nature the brakes were applied simultaneously on all vehicles and there was far less chance of the train 'snatching' as with the automatic air system, so smoother stopping could be consistently achieved. The system had one drawback as the EP brake alone did not satisfy the requirements of the 1889 Regulation of Railways Act which stated all trains had to have an automatic fail-safe brake. The driving position was fitted with a brake controller that the motorman could operate either brake from. Under normal circumstances this would have been the smoother and simpler EP brake but the train could quite comfortably be driven just using only the air brake if the EP system developed a fault. The automatic air brake had been successfully used alone on both previous builds of NER electric stock so the EP brake was considered an enhanced feature. Air was also used for the sanding equipment fitted in the driving ends and was operated by a foot pedal valve on the floor of the driving cab; this delivered a quantity of sand to the motored axles to cope with slippery rails caused by leaves on the line or heavy rainfall. As with previous stock a whistle was mounted at the driving end to the outer side of the driver's window and operated by a valve in the cab. Air operated window wipers were also fitted to each driver's window.

The motor coaches had electrical control equipment provided by Westinghouse. This was of the electro pneumatic type whereby line voltage contactors were opened and closed by air operated valves, an improvement on the electromagnetic switches of previous stock as they were less prone to rogue arcing. The cab fronts carried two destination blinds above the central access door, and these allowed the destination to be shown in the top blind and the type of service, such as STOPPING or EXPRESS in the lower window. In addition a four light cluster was mounted under the non driving side cab window which would be illuminated in various combinations to indicate the train and route. A single tail lamp was mounted above that same window that was illuminated to mark the rear of any train formation. Provision was still made for the carrying of external lamps if needed, with one bracket mounted to the lower cab front above where the buffers would have been. It should be noted that only the four single cars were built with buffers, and none of the articulated sets were so fitted. Below the sole bar on the front of each unit either side of the Cowhead coupler were the multiple working electrical jumpers and receptacles: three jumpers on the driver's side and three receptacles on the non-driving side.

The saloons were heated and lit by electricity, and large body side windows were provided with sliding top light ventilators above should additional air flow inside the saloon be required on a warm day.

A Metro-Cammell built 1937 Tyneside unit. The livery appears to be the post Second World War blue and cream. (British Rail)

An eight-car train of 1937 stock, wearing LNER livery on a coastal circle working. (Newcastle Central Libraries)

Interior of the 1937 stock as built with the 'bucket' seats. (Author's Collection)

The passenger accommodation was open plan inside with each car sub-divided into two semi-saloons and all internal walls were covered in cream or brown rexine. It was with the internal seating area that variations were apparent due to the need to provide various types and classes of seating for the full service. Unlike previous stock the diagram number was allocated to the set rather than the specific vehicle; given that it was impossible for the vehicles in an articulated set to be split and operated independently this made perfect sense. The Type A units were Diagram 234 and twelve in number. They were made up of a combination of Driving Motor Third

Articulated detail of a 1937 unit showing the electrical connections between the vehicle bodies. (ARPT)

car paired with a Driving Trailer Third with both vehicles having sixty-four seats.

The sixteen Type B units, Diagram 233, were formed of a Driving Luggage Motor Third and a Driving Trailer Composite. The luggage compartments had a double set of sliding doors for access from the platform; the driving car had fifty-two seats whilst the trailer had twenty-eight first and thirty-two third-class seats. It is worth noting that the third-class on the trailer car was considered 'convertible' accommodation, meaning it could be upgraded to first on services that had a higher demand for that class of accommodation. This was achieved by simply changing the accommodation classification digit held in a frame on the outside of the car by flipping it over from a '3' to a '1'; the actual seats were the same, just temporarily more exclusive. Seats in the saloons of all types of unit were of the bucket type that were also to be fitted to the South Tyneside units upon refurbishment, and there were a few tip up seats fitted in the vestibules for use when trains were crowded.

Eighteen Type C sets were built to Diagram 235 and were made up of a Driving Motor Third seating sixty-four and a Trailer Third seating sixty-eight. The final variation was the Type D set of which eighteen were built to Diagram 243. These had a Driving Luggage Motor Third and a Trailer Composite, seating fifty-two in the motor car and twenty-eight first and thirty-six third-class 'convertible' seats in the trailer. The more eagle eyed reader will have noticed that types A and C were virtually the same, as were B and D. This was true as the only real difference was the lack of a driving cab on the trailer cars of the Type C and D units. This made them less operationally flexible, meaning they had to be coupled back to back with driving cabs facing outward or the non driving end had to be coupled to a double ended unit of Type A or B. All vehicles had end doors fitted but they were for crew use only and passengers were not permitted to use them to pass between vehicles. The numbering system for the individual vehicles was based upon them never operating as such and being permanently coupled in pairs. The motor coach had an odd number and its trailer partner was one number higher. The table gives the sequence for the four types.

Table 5: Two-car sets delivered in 1937.

Unit Type	Diagram	Running Numbers
A	234	24145–24168
B	233	24169–24200
C	235	24201–24236
D	243	24237–24272

Thus the first-built Type A unit was formed of Driving Motor Third 24145 coupled to Driving Trailer Third 24146 and the last built Type D unit was formed of Driving Motor Luggage Third 24271 coupled to Trailer Composite 24272. The articulated sets were completed and delivered to Tyneside in 1937.

Two other passenger vehicles were built that were not articulated, these were two Driving Motor Luggage Thirds with driving cabs at each end and fitted with two motor bogies and conventional buffers in addition to the Cowhead coupler. These two cars were constructed to Diagram 280 and were slightly longer than the articulated cars, with a body length of 59 feet. In addition to the luggage compartment they had two passenger saloons with a combined seating capacity of fifty-two. It was these two vehicles that became the preferred choice of motive power for the six-car 'Controlled Set': one vehicle was marshalled either end of the formation, hence the conventional buffers. The Controlled Set by this time was formed of six-bogie carriages and these had to be modified with the same electrical multiple working connections as the new stock. Although they had EP brakes fitted, they still lacked any form of heating but at least the lights worked. The train earned the nickname of 'The Pullman'. Parcels vans had been employed as motive power for this train in its earliest incarnation and still appeared from time to time under the LNER, but the single cars gave the train a full eight coaches of passenger accommodation, not to mention heating in the Driving Motor Thirds. What the old compartment coaches lacked in comfort they made up for in having a very high passenger capacity, which was useful as the workmen's trains on the Riverside branch this set operated were extremely busy. The set was utilised on two trips Monday to Friday and a single one on Saturday, while the rest of its time was spent at Gosforth car sheds. It could be brought into action if needed during this down time but no evidence exists of it ever being so. The single cars were fully compatible with the rest of the articulated fleet and could, if required, be coupled and operated within that fleet. The final two vehicles were two Motor Luggage Vans built to Diagram 277, which had the same dimensions and traction equipment as the two Driving Motor Luggage Thirds and also sported conventional buffers and draw gear. Like previous builds of such vans for the Tyneside electric network they had two internal compartments: one for parcels and one for fish, the former just over 27 feet in length and the latter just over 24 feet long. As they were required to haul steam hauled parcels vans these two vehicles were fitted with vacuum exhausters so they could additionally control a vacuum brake as used on that type of stock, meaning they had a total of three braking systems on a single vehicle. These four single vehicles were individually authorised to haul a trailing load up to 100 tons, but if running on their own, or with a load of less than 30 tons, then the motors on one of the bogies had to be cut out.

24178 in LNER livery at Newcastle Central sometime between 1948 and the unit renumbering of 1951. (Author's Collection)

Bucket seats, flat caps and furry hats. An everyday scene of travellers on the electric network in February 1967. (Colour Rail)

The whole fleet, including the controlled set coaches, was painted in the red and cream livery that would also adorn the refurbished South Tyneside units: red lower body side and off white upper body with lettering and numbers in gold, harking back to the days of the earliest NER electric service.

Before the new stock entered service, the position of the conductor rail had to be altered on the North Tyneside lines to conform the new national standards which had been put in place by the Railway Electrification Committee, chaired by Sir John Pringle. Their report was published in July 1928 with a view to allow for stock to be interchangeable and where possible through running to be achieved. The recommendations also encompassed a standard supply voltage of up to 750v DC for top contact conductor rails. Standards of conformity were also set by the report for lines electrified by overhead wires. In order to conform to the new standard, although the report made clear it was only a recommendation, the LNER had to move its conductor rails on the 72 miles of the whole network, including depots and sidings, from the original position of 1 foot 7 1/4 inches from the centreline of the conductor rail to the inner edge of the nearest running rail, to a new position. The report stipulated the conductor rail centre line had to be 1 foot 4 inches from the inner edge of the nearest running rail. The height of the top surface of the conductor rail was to be 3 1/4 inches above the top of the nearest running rail. The work was to be undertaken over four days, in phases so as to minimise the disruption to the service. On Friday 22 May 1936 the Riverside branch was switched off and the service was replaced by steam trains whilst gangers slewed the conductor rails into their new position. In the following two days the work continued on all but the main circular route, which had its turn to be switched off and substituted by steam trains on Sunday and Monday 24 and 25 May 1936. The mammoth task was completed by eight gangs of workmen working round the clock shifts. One can only imagine the degree of planning that went in beforehand and once the work was in progress on the ground. Whilst a double track section of line with lengthy runs of conductor rail may appear a simple enough task on paper, one has to consider that all the short sections that adjoin point work at busy locations such as Newcastle Central station and South Gosforth car sheds were numerous. It has to be admired that those who both planned and executed the scheme had completed the work by teatime on the Monday, so the system could be successfully tested at 9pm that night. The stock had its collector shoes modified to fit the altered conductor rail position during the mass switch off and the whole project was completed in time for the full electric service to resume on all lines at the start of Tuesday 26 May. In modern times it is common for lengthy possessions to be taken over bank holiday periods and a chunk of the network closed, buses and coaches employed to ferry customers and much grumbling to take place; over the Whitsun bank holiday of 1936 on Tyneside all lines remained open, albeit with steam instead of electric traction, the trains ran and there were no over running engineering works when people came back to work.

The first of the new Met-Cam units entered passenger service in July 1937, very quickly followed by the remainder of the fleet. They were an instant success with passengers who appreciated the new style over the tired look of the original fleet. Journey times were also improved: a whole ten minutes was shaved of the stopping service travelling out and back from and to Newcastle Central via the coastal loop. Eventually all the original 1904–1915 built first generation of Tyneside electric stock were taken out of traffic and dispatched for breaking up at various locations. There were a few survivors of the original stock into the next decade: fifteen passenger vehicles were retained for use on workmen's trains up until 1940 and then these were placed into store as standby units during the Second World War in case an excessive number of units be lost to enemy action. The vehicles were moved away from Tyneside and stored at various locations, including Darlington, Thirsk, Scarborough and Staveley. They were never needed and were quietly scrapped after the end of the war.

The two 1904-built Motor Luggage Vans found further use from 1938 on the network in the guise of de-icing vans. For this purpose they lost their traction motors and control gear and had tanks fitted in one of the former luggage compartment that contained a kill frost solution. This was delivered to the top surface of the conductor rail via pumps and pipework that fed onto the special collector shoes on the shoe beams on the bogie at that end of the car. The way this was achieved was to tow the now motorless van around the network at night with a steam locomotive. The delivery system fitted to these two vans was quite ingenious. The conductor rail changes sides at various locations around the network, so to avoid the van continuously spraying where it wasn't needed the delivery head was fitted with a valve that was opened and closed by the actions of a lever that itself was raised and lowered by a rubber wheel that traced the position of the conductor rail. When there was a conductor rail present the wheel rode along it and held the arm up, which opened the valve and enabled the de-icing solution to be sprayed that side. When there was no conductor rail the wheel was down which lowered the arm and closed the valve. Both of these vans continued in this role until finally laid aside in 1966. The 1908 built motor luggage van was also transferred to departmental duties in 1939 but was laid aside by 1955.

The outbreak of the Second World War caused a reduction in services as a wartime economy. It was also responsible for a

New Trains, Old Routes • 63

West Monkseaton station building pictured in September 2019. The art deco lines built by the LNER have survived to see use as part of the Tyne & Wear Metro.
(Graeme Gleaves)

On 13 January 1963 one of the converted NER parcels vans was parked at a snowy Gateshead.
(Colour Rail)

Bomb damage to NER South Tyneside EMU car No. 23249 at South Shields after a night raid in May 1941. (LNER/RAS Collection)

significant visual change. The RAF told the LNER that the red and cream colour scheme of their electric stock made them very visible from the air and thus easy targets for enemy air attacks. The LNER responded by changing the livery of the whole fleet north and south of the river to one of Marlborough Blue below the waistline with Quaker Grey above it. Units were hastily repainted from 1941 and lettering in the corporate Gills Sans font was applied along with the LNER logo in 'lozenge' totem form at the centre of the body side. Whilst the standby units were never required there were vehicles

lost to enemy action: one complete Type C unit (24229 + 24230) of Met-Cam stock was damaged beyond repair in 1941 and scrapped with no replacement set ever built. Also in May 1941 one of the 1920-built replacement cars (23249) was destroyed in an air raid at South Shield; again no replacement for it was ever built.

By the end of the war the nation's railway system had once again been pushed to the limit by attacks by the enemy but mainly by the huge demands placed upon it from supporting the war effort. Once again there would need to be a significant change to take the industry forward. The LNER made

Photographed in December 1948, a rare colour image of a Met-Cam unit in LNER blue and cream at Newcastle Central. (Colour Rail)

One of the Metro-Cammell LNER Motor Parcels Vans at Newcastle Central. The blue and cream livery was introduced after the RAF claimed the previous red and cream livery was too visible to enemy aircraft. (Colour Rail)

LNER North Tyneside Electric Parcels car E29467E in malachite green at Gosforth car sheds in September 1960. This had been LNER 2424. (Colour Rail) one final contribution to the network when a new station was opened at Longbenton, situated between South Gosforth and Benton. The station opened on 14 July 1947 and its construction was brought about by the opening of new government offices for the Ministry of Pensions and National Insurance at that location. The station was only to be served by a limited number of trains Monday to Friday, to coincide with office hours, and no extra time was allowed in the timetable for the station stop. Over time, long after the passing of the LNER, Longbenton became a full-time station with trains during the week and weekend to serve the new housing estates that sprung up nearby. The station was also 'renamed' under British Railways to Long Benton.

Chapter 9

NATIONALISATION, BRITISH RAILWAYS

Twenty-five years to the day after it came into being, the LNER, like the NER before it, passed into history. It was replaced by British Railways, a state owned company that took over the running of all the railways, along with some of the sea and road concerns of the 'big four'. The Labour government that had been elected in 1945 under Prime Minister Clement Atlee had promised to bring major industry under state control. This included not only railways but other large industries such as coal mining, which became nationalised in 1946, followed by power stations in 1947 with the formation of the Central Electricity Generating Board, whilst local gas supplies would be nationalised in 1949. In 1948 the focus was on the creation of both the National Health Service offering free health care to all and launching the British Transport Commission (BTC) to provide a nationally owned and more integrated transport network. The Transport Act of 1947 had created the BTC which would oversee nearly all inland transport, be it rail, canal or road haulage. It should be noted however that there were still some elements of each sector that remained independent private concerns or under the control of another government or local authority body. One exception was the London Underground, which was run by the London Passenger Transport Board. The LPTB came under the control of the BTC despite already being publicly owned. The nationalised sectors were grouped under the British Transport Commission who in turn set up executives to run the various aspects of the transport system. British Railways was one of those executives and took over the network of both the big four and several other minor railway companies at the stroke of midnight on 1 January 1948.

On Tyneside these changes became progressively visible over the next few months as the LNER identity was replaced with the new corporate one for the nation's railway. On stations the LNER posters were replaced by British Railways ones and the station name boards changed to a shade of orange, or tangerine, with white lettering, Staff uniforms were changed to a standard issue that was the same in Scotland, Wales or anywhere else that British Railways were in charge. The most visible change started appearing in 1949 when the LNER's blue and grey colour scheme gave way to all-over British Railways green on the fleet of EMUs. The company's crest was carried on the motor coaches but not the trailers; underframes remained black and roofs stayed grey but the lettering and numbering went from gold to a pale cream colour and was no longer in the LNER's Gil Sans font. Car numbers were prefixed with the letter E to denote that the owning region was the Eastern, but it was not that simple as Tyneside fell within the North Eastern Region of British Railways and was run from offices in York, much as it had been under both the LNER and NER. The spread of the BR identity was gradually rolled out but the service remained pretty much as it had done during the latter years of the LNER.

Changes and improvements were not far away, commencing with the renumbering of the whole electric fleet north and south of the river

A four-car formation of 1937 stock forms the 12:35 Newcastle to the coast service on 28 May 1962, pictured at Heaton station. (Michael Mensing)

in 1951. This was brought about due to the extensive construction of new Mark 1 coaching stock that required the number bracket currently occupied by some of the Tyneside electric cars, and as no two vehicles could run on British Railways with the same number. The older stock was renumbered to make way for the new builds to use the required sequence. The renumbering of the 1920 fire replacement stock that operated the South Shields services allocated numbers starting at 29175 upwards for the Driving Motor Coaches and reached 29192. Eighteen numbers were allocated despite the fact that one car had been lost during the war and had not been replaced. This was originally 23249 and was allocated 29188 under the new scheme, but the number was of course never used. The cars were not renumbered in build order but in a sequence that had the Driving Motor Thirds at the lower numbers and the Driving

A busy day at Tynemouth station, possibly at a weekend given the number of children who have got off the train. The unusual white cross on the lamp iron is not a feature seen in any other photos.
(Peter Shoesmith)

Motor Luggage Composites in the upper half of the sequence. The trailers were re-numbered from 29375 up to and including 29392 as all 18 were still extant. The 1920 Motor Luggage Van 23795 was renumbered to 29493. Renumbering the Metro-Cammell built cars was no less complex, as mentioned in a previous chapter: the articulated sets had consecutive numbers with the motor half having the lower odd number and the corresponding trailer being one number higher. British Railways' 1951 scheme saw all motor cars numbered from 29101 for the first-built A type unit, then each subsequent motor car carried a consecutive number until 29164 for the last-built D typeunit. The trailers used a different approach beginning with 29301 and going all the way up to 29328 for the last built B type unit where things got a little complicated as the first-built C type trailer was renumbered 29229 then numbers then ran consecutively up to 29264 for the last-build D type unit. In order to maintain the build order continuity this scheme even allocated the numbers 29143 and 29243 to the C type unit (motor 24229 and trailer 24330) that had been destroyed by an air raid in 1941 and had subsequently been scrapped that same year. Just like the missing 1920 car, the numbers were allocated but never used. The two double Driving Motor Luggage Thirds 24273 and 24274 were renumbered to 29165 and 29166 respectively, whilst the two Motor Luggage Vans 2424 and 2425 became 29467 and 29468 under the scheme.

Table 6: British Railways renumbering scheme.

| South Tyneside Stock Renumbering ||||
| Driving Motor Cars || Driving Trailer Cars ||
Old Number	New Number	Old Number	New Number
23212	29175	23180	29375
23219	29176	23195	29376
23221	29177	23197	29377
23225	29178	23198	29378
23226	29179	23204	29379
23227	29180	23211	29380
23228	29181	23507	29381
23238	29184	23508	29382
23241	29185	23512	29383
23243	29186	23513	29384
23246	29187	23518	29385
23249 (vehicle scrapped)	29188 (never carried)	23785	29386
23250	29189	23788	29387
23251	29190	23790	29388
23253	29191	23791	29389
23520	29182	23217	29390
23521	29183	23268	29391
24465	29192	23524	29392

| North Tyneside Stock Renumbering ||||
Old Number Range (odd)	New Number Range	Old Number Range (even)	New Number Range
A Type Driving Motor Thirds		A Type Driving Trailer Thirds	
24145–24167	29101–29112	24146–24168	29301–29312
B Type Driving Motor Luggage Thirds		B Type Driving Trailer Composites	
24169–24199	29113–29128	24170–24200	29313–29328
C Type Driving Motor Thirds		C Type Trailer Thirds	
24201–24235	29129–29146	24202–24236	29229–29246
D Type Driving Motor Luggage Thirds		D Type Trailer Composites	
24237–24271	29147–29164	24238–24272	29247–29264

Since 1941 the South Tyneside fleet had been left with a spare driving trailer following the loss of one of the driving motor cars during the war. Whilst the fleet did not operate any fixed formations and motors were paired with different trailers as maintenance and servicing needs dictated, this was not a specific car that was left spare but just a numerical imbalance. In 1951 a conversion was carried out on Driving Trailer Third 29388 that left it unable to work with the remainder of the 1920 stock and saw it re-allocated to duties north of the river. It became Britain's first ever Perambulator Van. The thinking behind the 'PramVan' was to create a vehicle with easy access and room to store the large coach-built prams that were commonly used at the time before more portable and compact buggies became the norm, and thus make access to travel easier for those wishing to take their infants on the train to the coastal resorts on the North Tyneside network at weekends. The conversion work undertaken in September 1951 involved the removal of the driving compartment and also the bucket seats from the interior saloon, which were replaced with bench seats along the side of the interior, this gave more space inside for prams to be manoeuvred. The double doors provided at the centre of each side ensured there was plenty of room to get the prams on and off from platform level. The electrical connections were altered to enable the vehicle to be marshalled between a pair of Met-Cam articulated units, essential as this Pram Van had no driving controls of its own.

In addition to this vehicle the two Met-Cam built Double Driving Motor Thirds (Diagram 280) were modified for perambulator van duties. They retained all driving controls and traction equipment and the modification was less drastic than that on the ex-NER vehicle. The prams could be loaded via the double doors of the luggage compartment and then moved into the adjacent saloon, as the door between that and the luggage compartment was removed, as were the seats in that area which were replaced with bench seats along the wall of the saloon. The revised layout had a seating capacity of

The 1937 stock did on rare occasions work services on the South Shields route. Here a six-car formation is crossing the High Level Bridge with such a service. (ARPT)

A four-car North Tyneside service heading for Newcastle at South Gosforth in an everyday scene in the late 1950s. (Author's Collection)

forty-eight. These vehicles retained full multiple working capability and could operate with the rest of their Met-Cam built brethren exactly as they had done so before.

A serious accident claimed a Met-Cam articulated set during 1951. On 17 August the 10:35 departure from Newcastle Central to the coast started out of Platform 2 despite the starting signal being held at danger due to a late running inbound working. Quite how this happened is a stunning example of human failure as it took no fewer than three different people to fail to see the starting signal was not cleared: the platform dispatch staff, the guard of the train and ultimately the driver. The result of this was that the outbound train collided with the inbound one just outside of the station. The trains met corner to corner and the leading vehicles of each set were seriously damaged. They were Driving Motor Luggage Third 29131 and Driving Trailer Third 29312. When the wreckage had been cleared both vehicles were scrapped and a new unit was formed from the undamaged halves of each, formed of Driving Motor Third 29112 and Trailer Third 29231. Because the surviving vehicles made this a C type unit, for operational reasons the driving motor was renumbered to 29131 to keep the set in the correct number range.

The Pram Van conversion had restored numerical balance to the South Tyneside fleet in 1951, but that was to be short lived when in January 1954 an accident that occurred at Gateshead caused Driving Motor Luggage Composite 29131 to be written off. The failure of a set of points at Gateshead High Street Junction caused the vehicle to attempt to take two separate routes as it entered Gateshead East station, with the rear bogie attempting to take the line to Greensfield Junction. The vehicle was at the rear of the formation and collided with the signal box at this location more or less side on. The damage to both signal box and carriage was considerable; the former was repaired but the railway vehicle was not and was officially condemned in May 1954. For those of you who believe in fate then there is a strange slice of it attached to this incident. The damaged car 29131 had been renumbered from 23253 and under the NER had been car number 3253, built in 1928 to replace the previous car number 3253 that had been destroyed in the 'driverless' crash at Manors on 7 August 1926; now if you add up all the digits of 3253 you get unlucky 13 and the accident at Gateshead occurred on 13 January. Just a co-incidence surely?

The loss of this car once again gave the South Tyneside routes an imbalance of driving motor and driving trailer cars, with a spare trailer being the result. It was not an issue that needed reacting to as plans were already in hand to replace the entire fleet of 1920 stock with brand new two-car units built to a BR standard design. The full rolling stock replacement would

An LNER North Tyneside EMU arrives at Percy Main bound for Newcastle on 29 August 1956. (Rail Archive Stephenson)

The 1:05pm service from Newcastle Central out via Jesmond to the coast arrives at Manors on 28 May 1962. (Michael Mensing)

be carried out by the end of spring 1955 but that was not to be quite the end of the fire-replacement NER cars. The solitary Pram Van conversion carried out in 1951 had proved both popular and a success, so with the withdrawal of the fleet planned, three of the Driving Trailer Thirds were set aside and converted to the same specification as the original Pram Van. The vehicles selected were 29376, 29387 and 29390. The conversions were carried out in 1955 after the BR-built stock had taken over the South Tyneside services, and so by the close of 1955 it was possible to view examples of four generations of electric stock still active on Tyneside. The new BR units ruled south of the river whilst the LNER Met-Cam units held court north of the river. In addition the four NER built 1920 Pram Vans were sandwiched into north Tyneside services, and during the winter months both the 1904 NER-built Motor Luggage Vans were used in their new guise as de-icing vans.

The Riverside branch had always existed as an aside from the main coastal loop services and this never changed under either the LNER or British Railways. The line had a basic hourly service and was patronised predominantly by workmen from the dozens of industrial sites located on the Tyne's north bank. The areas around the stations on this branch (in reality a loop line) were never going to become centres of residential traffic and thus the service was always busiest at around the time of shift changes, with one set of workers arriving to take over from those who would then be going home. Byker station, the first one on the line after the junction with the coastal circle route, was very close in proximity to Heaton station on the coastal route which enjoyed a vastly superior service. Unsurprisingly Byker never developed a heavy patronage and when British Railways looked to make savings Byker station was closed on 5 April 1954.

An empty coaching stock working from South Gosforth shed to Newcastle Central passes through Manors on 21 May 1962. (Michael Mensing)

Another four-car formation at Newcastle Central in the mid 1960s. The small yellow warning panel has been added to the front. (R. Carroll Collection)

E29103E leads a four-car service on the North Tyneside line near Benton where it is about to cross above the East Coast main line. (R. Carroll Collection)

A highly unusual working caught on camera. One of the 1937 Motor Parcels vans is hauling a Met-Cam built Class 111 DMU buffet car, NE59577. The train is pictured heading towards Benton. (R. Carroll Collection)

Another view of the same working with the parcels van using its vacuum exhauster to operate the brakes on the DMU car. The reason for the movement of the buffet is unknown but it is probably heading to South Gosforth depot. (R. Carroll Collection)

A view of Motor Parcels van E29468E working solo through Manors. There were regular parcels workings and often these involved the vans hauling a trailing load. The image is dated 28 May 1962. (Michael Mensing)

A pair of the 'Pram Van' vehicles converted from 1920 stock, pictured in store at South Gosforth in the very early 1960s. (Colour Rail)

Chapter 10
BRITISH RAILWAYS ELECTRIC STOCK

The final third rail passenger stock to be built for the Tyneside network was constructed through 1954 and delivered during early 1955. The original South Tyneside electric service had seen eighteen two-car units of NER-built 1920 stock provided. This figure had been based more upon what was available rather than what was needed, as the 1920 stock offered a self contained subclass that could be segregated out of the wider Tyneside fleet and utilised as a dedicated fleet on this route. The service had not suffered when one unit was lost in 1941 and then another in 1954 so it should have come as no surprise that the new fleet for the route would be even smaller than the previous one and comprise just fifteen two-car units.

These were built under the British Railways stock design where the emphasis was on standardisation: what worked in Cornwall also worked in both Scotland and Surrey, or so the basic theory went. The thinking behind standardisation was that parts could be cheaper to make as they were mass produced, and spares would be interchangeable between fleets. Staff would be familiar with operating and maintaining stock regardless of their location and this would make training easier too. Within this 'standard' structure there was some scope to tweak the designs to meet local criteria. In the case of passenger stock British Railways had developed the Mark 1 coach and that came in two types: express and suburban. The express type was gangwayed with just three doors per side and had seating in either bays of four or four-a-side compartments. The suburban was the same shape but laid out to meet the needs of rapid loading and unloading of commuter trains, with doors to every seating bay, except in first class, which had side compartments and the stock was non-gangwayed and featured a mix of full width compartments and semi-saloons. The fifteen units being built for the South Tyneside route were suburban type carriages and were an add on for a fleet of seventy-nine units being built for service on the commuter routes from London to North Kent. Each unit consisted of two coaches: the Driving Motor Brake and a Driving Trailer Composite, and were built by the former Southern Railway carriage works at Eastleigh.

Front end view of a newly built South Tyneside EMU just after it was rolled out of Eastleigh Works. (British Rail)

A work-worn unit pictured at a wet South Shields station. (SERA Archive)

The bodies of each vehicle were 63 feet 11½ inches long and sat on an all steel frame. The body and underframe made extensive use of welded construction and were pressed and assembled on construction jigs that were the same as for the standard suburban steam stock. Single bolster bogies using riveted sub assemblies were provided with 3 foot 6 inch diameter wheels for the trailers and 3 foot 4 inch wheels on the single motor bogie; this was mounted under the guard and luggage van end of the unit and featured two 250hp English Electric EE507 traction motors. Primary suspension was by leaf springs above each axle, with nine leaves in the springs for the heavier powered ones and seven for the unpowered axles. Between the bogies of the Driving Motor Brake car and mounted to the underframe were the various boxes containing the electro-pneumatic contactors for control of the traction and heating as well as for the Westinghouse DH25 air compressor. A battery box was also provided with cells arranged to give a 70v DC output to power the control and emergency lighting, although these batteries could not power the train's motors or heating. Under normal running conditions the heating, which was mounted in the spaces beneath the seats, was fed from the third rail supply. The lighting was powered by the 70v DC output of a motor-generator set that was also mounted on the underframe of the motor car and ran constantly if the unit was connected to a live third rail. Lighting was provided by tungsten bulbs mounted in two separate lines, one on either side of the roof centreline. As with

Publicity shot of the interior of the first-class compartment of the Driving Trailer of a BR South Tyneside unit. (British Rail)

previous stock the trains had two braking systems: the automatic air brake was the fail-safe brake and the electro-pneumatic brake was the normal service one. The EP system was slightly more advanced than that fitted to the Met-Cam units as the system could self lap on the Eastleigh built units and thus needed only five positions on the driver's brake controller to control both brakes. Above the floor seating was provided for seventy-four third-class passengers in the Driving Motor Brake and ninety third-class and eight first-class in the Driving Trailer Composite.

The provision of a single first-class compartment was the first variation on the Tyneside units from those built for the Southern Region, where commuter traffic was all one class. The third-class seating was arranged in two semi-saloons on the motor coach with 6-seat benches at each bulkhead and 2+3 seating in the area between, The Driving Trailer Composite had one such semi-saloon and four full width third-class compartments in addition to the solitary first-class one. Each bulkhead was fitted with a central mirror and flanked on either side by a picture frame that featured reproductions of watercolour images of regional beauty spots. Above these were the luggage racks strung with netting, while the first-class compartment had the added luxury of a smaller rack mounted below for the placement of passenger's umbrellas. As these were suburban units there was a hinged door to every seating bay to facilitate the rapid loading and unloading at platforms during the morning and evening rush hours. The passenger doors were identical to those fitted to other Mark 1 carriages and had spring loaded door locks that could be operated from inside or out, along with a Beclawat droplight window. The doors were supported and swung on three self-aligning, ball seat hinges.

Current collection was once more by gravity shoes mounted on a wooded beam, one either side of the outer bogies. The Eastleigh-built units were the first fleet provided for use on Tyneside that did not have the Cowhead coupler; instead the preferred choice of a 'buckeye' drop-head automatic coupler with retractable side buffers was fitted at each cab end. The buffers would only need to be extended and used in the event of the unit being coupled to a conventional drawgear fitted loco or train in the event of a failure. Between the two cars a semi-permanent three link coupler was provided with a single central buffer. These units were meant to stay in a fixed pair,

Floor plan and dimensions of both vehicles of a BR South Tyneside unit. (Author's Collection)

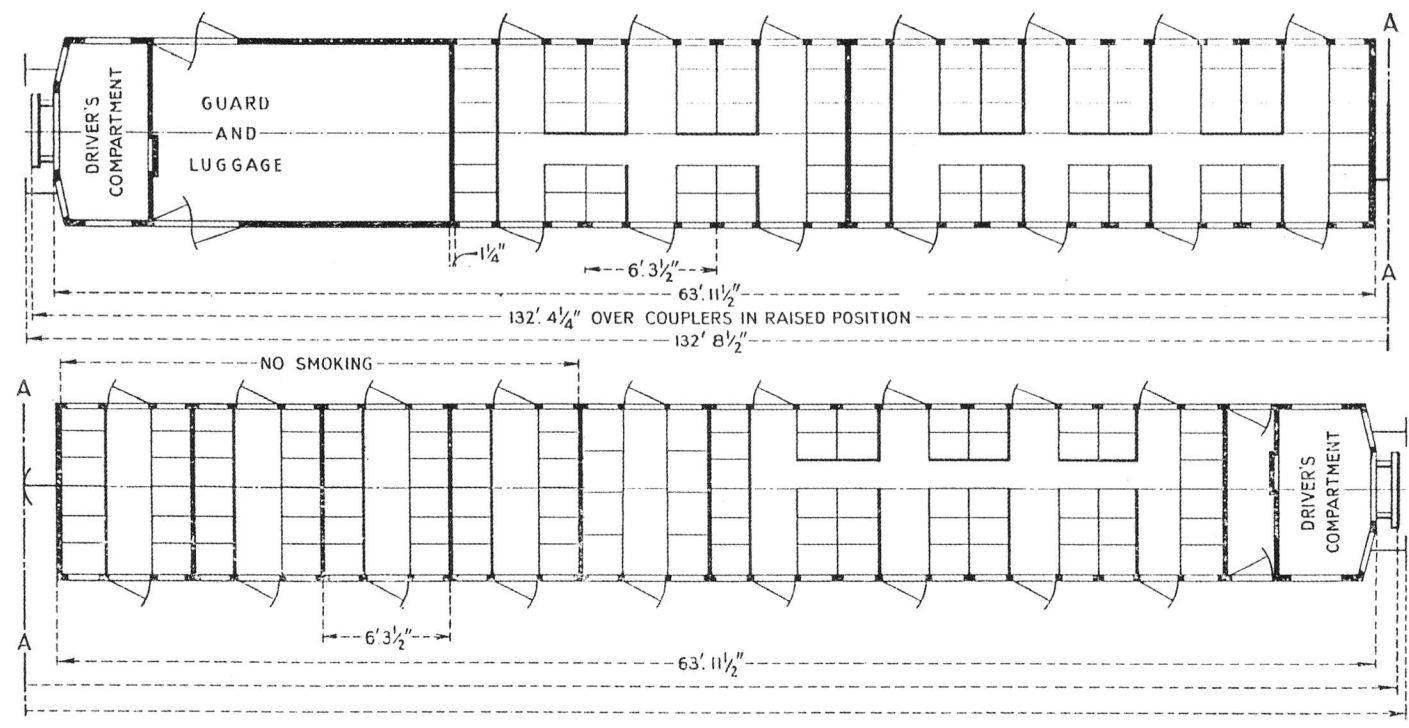

unlike previous NER stock and the coupling reflected that. Units could be coupled together in multiple, and air and electrical connections were provided at waist height either side of the cab front below the cab front windows, a departure from previous Tyneside stock where the electrical and air connections were mounted below the buffer beam at each end. The Eastleigh arrangement made coupling and uncoupling much easier at stations but proved to be far more awkward in depots and sidings as the shunter had to climb up between the vehicles and perch on the small steps, or Pullman rubbing plate, whilst the air and electrical connections were joined.

A full width driving cab was provided at each end which on the Driving Motor Brake was accessed via the brake van, and on the Driving Trailer there was a vestibule with inward opening doors to allow the driver to get in. The guard's brake was the next big difference between the units built for Tyneside and those for service in the south. The need to convey boxes of fish, prams, parcels and other luggage on the South Shields route called for a much larger luggage space, over 15 feet long. At the rear of which, the opposite end to the driving cab, the guard was provided with a swivelling bucket seat and two periscope lookouts, one for each direction of travel. The final physical difference from the Southern Region units was the provision of the headcode lights and destination blinds on each cab front between the windows. At driver's eye height was the cluster of four white lights, the same as had been provided on the Met-Cam stock, but the rectangular plate they were mounted on had the red tail lamp lens placed in the centre of the marker lamps. Above that was a roller destination blind with two blinds: the top one stated the destination whilst the lower one gave the type of train, such as STOPPING or EXPRESS. Destinations for both north and south of the Tyne were provided, as even though these units were built for the purpose of taking over the South Shields line it was recognised they could be utilised on other duties. The pneumatic whistle was mounted above the driver's window and operated by compressed air, which also operated the window wiper fitted to that same window. Two other differences between the Southern and Tyneside built units worth mentioning are, firstly that the seat cushions on the Tyneside units were of the type fitted to the steam hauled suburban Mark 1s, whilst the Southern ones had a different style of seat assembly. The reason for this may be that the steam-hauled variant of the Mark 1 suburban was in use on the Eastern Region routes out of Kings Cross and Moorgate, so to keep a spare parts pool that was suitable for all suburban stock the Tyneside units were fitted with the loco-hauled seating. The other difference, and one that was impossible to spot with the naked eye, was that the Tyneside units were geared for a maximum speed of 90mph as opposed to the 75mph of the Southern units. The reasons for this change are unclear as the start/stop nature of the Tyneside network would have meant the units had no realistic opportunities to travel that fast, and no stock built prior to this had ever been geared to go above 60mph, so why this was part of the design remains a mystery.

The units were finished in the British Railways standard EMU dark olive green and had all lettering in cream. This was limited to just the vehicle numbers, the large

Another view of the High Level Bridge over the River Tyne from the Gateshead side and with a pair of BR South Tyneside electric multiple units making the crossing on 26 June 1960. (Ron Fisher)

A six-car train forming a Newcastle to South Shields service crosses the High Level Bridge and approaches Gateshead East on 26 June 1960. (Ron Fisher)

number 1 that adorned the exterior door of the solitary first-class compartment, and 'Guard' on the inward opening door to the brake van, whilst the inward opening door behind the cab of the driving trailer had 'Private' on it. The circular British Railways emblem was applied only to the motor car, half way down each body side. The units would go on to be numbered consecutively from the Southern Region builds, but unlike them the Tyneside units had no unit number and were identified only with vehicle numbers. The units were formed into their two-car sets in build and numerical order, with the Driving Motor Brake vehicles being numbered from E65311 up to E65325 and the Driving Trailer Composites being the sequence from E77100 up to E77114. Completed units were out-shopped from Eastleigh works during late 1954 through early 1955. They were towed north to their new home depot at South Gosforth car sheds where the shoe gear was fitted and the units were tested and commissioned ready for traffic. Staff training was also undertaken on the network. The first examples started to appear in passenger traffic in March 1955, and on 17 May 1955 the last examples of the 1920 fire replacement stock ran in passenger service on the South Shields line. After that date the only examples in use were the four Pram Vans on the north side of the river. The 1920 motor luggage van remained in traffic for another year as its replacement was still under construction. The remainder of the 1920 stock went into store at South Gosforth and a few other locations until it was all quietly towed away and scrapped.

Eastleigh works were tasked with building what would turn out to be the very last third rail electric vehicle supplied to the Tyneside network, although no one knew that at the time. The Motor Luggage

Three-quarter publicity still of newly completed South Tyneside Motor Luggage Van E68000. (British Rail)

Side view of the then brand new E68000. The three banks of traction resistance grids are clearly visible under the brake van area. These were twice the size of those on the motor coaches of the passenger units due to the second power bogie. (British Rail)

Van being built to complement the new South Tyneside stock was to replace the 1920-built van 29493, and by April 1956 it had been put into traffic. The older vehicle was sent away for scrap as there was to be no second life as a de-icing van for that veteran, unlike its 1904 predecessors. The new van was numbered E68000 and followed the outline of the BR-built two-cars with its Mark 1 coach profile. The van was essentially a loco hauled full brake but with the EMU cabs added to the end. The body was 64 feet and 5 inches long over headstocks. The cab fronts were identical to the Eastleigh-built sets but it had internal features found on earlier Tyneside Motor Luggage Vans. It had two power bogies with the same traction and control equipment as the BR sets and could be coupled in multiple with them, and it was also provided with a vacuum exhauster to allow it to control a vacuum brake and haul vans.

Internally there were two compartments, one of 27 feet in length and the other of 21 feet 6 inches. Between these was the guard's compartment complete with periscope lookouts and swivelling bucket seat. The vacuum exhauster and chambers were mounted to

Internal view of the smaller compartment of E68000 looking towards the brake van with the Northey exhauster in the cages below the vacuum reservoir cylinder. (British Rail)

Floor plan layout and dimensions of E68000. (Author's Collection)

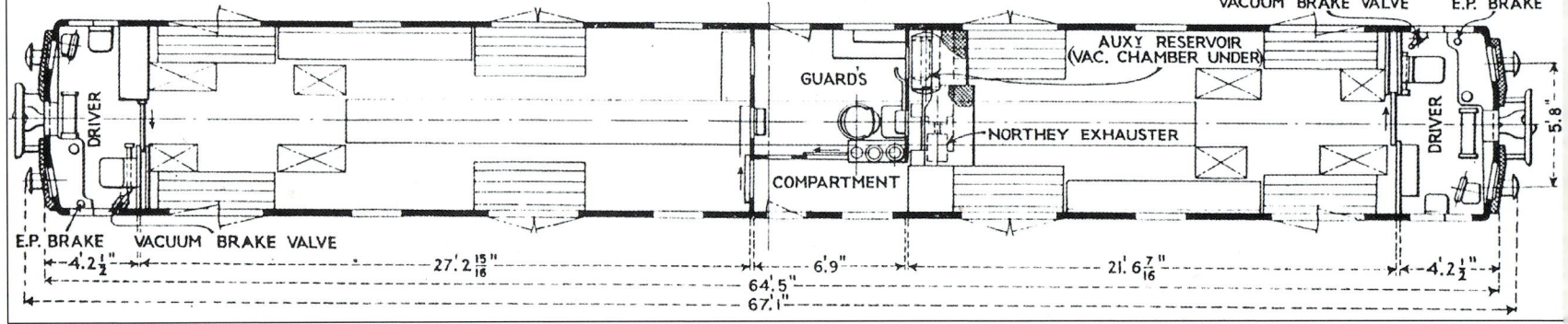

the external wall of the guard's compartment facing into the shorter of the two internal compartments. Externally the guard had a single leaf inward opening door but along the body either side of the central guard's door were fitted two pairs of double doors for the loading of parcels and fish traffic. The doors either side of the guard's one were both outward opening, whilst those behind the cab had one inward and one outward door, the same as the luggage vans on the passenger units. Tip-up benches were fitted to the walls of the internal luggage spaces between the doors so the van could be used to convey workmen. Buffers and drop head buckeye couplings were provided at each end, the same as those on the Eastleigh built two-car units.

Prior to delivery to Tyneside the unit was tested on the Southern Region, hauling goods vehicles and coaches. These trials were conducted on the Brighton main line with the vehicle reaching both London Bridge and Brighton under its own power. The van had to go back to Eastleigh Works on at least one occasion for rectification to some problem areas that had been revealed during its commissioning trials down south, but by March 1956 it was towed north to South Gosforth and thus began its service life on Tyneside. As mentioned this was to be last vehicle built for the original third rail network; the times were changing and the railways had not kept pace. Something had to give.

Internal view of the larger compartment of E68000 looking from the brake van towards the cab. The tip-up bench on the wall is in the raised position. (British Rail)

A postcard showing E68000 arriving at Brighton on a test run prior to movement north to Tyneside. (Pamlin Prints/Author's Collection)

South Tyneside unit formed of E65316 and E77105 waiting to leave Platform 6 at Newcastle Central with the 4:03pm for South Shields on 19 May 1962. (Michael Mensing)

Chapter 11
THE END

The summer of 1955 would have given the casual observer the impression that the future was bright on Tyneside's electric network. The new stock was in place on the South Shields line, British Railways' corporate image was everywhere, whilst nationally the business was embarking on its much lauded modernisation plan that aimed to bring the network out of the gloomy years of post-war austerity and into a bright future by phasing out the old and introducing new technology. This meant scrapping steam and ordering new diesel locomotives and railcars whilst embarking on main line electrification projects. None of this would come cheaply and thus British Railways was required to borrow heavily from the government purse. Despite being state owned it did not have a blank chequebook and had to repay what it spent; unfortunately the expectations and the reality were not going to align and before too long the reality of a serious rot began to set in. The problem was that with the ending of rationing, notably that of petrol, the average British family had aspirations to seek more status. This was reflected in the quality of the homes they lived in, the appliances that graced them and most significantly the cars that were parked in the driveways. This new air of optimism led to a desire for freedom and to spend better quality leisure time. Tyneside was no different to any other part of the country; the new class of professional who aspired to greater choices and expectations than their parents had had in life was on the rise. Public transport, whilst it had its place, was not high up the list of priorities. A man who drove himself to work during the week and his wife and children to the coast at weekends considered himself a success in this brave new era for Britain. Patronage of the railways had been in decline for several decades and whilst the effects of the economic depression followed by the Second World War had contributed to this, no amount of modernisation was going to reverse what was a significant cultural change. However, in 1954 British Railways did not see it this way and embarked on a scheme of modernisation to make the railways a part of this new post-war age. They estimated that by modernising the network it would become more popular, patronage would rise and thus revenue, and the debt could be repaid. When the white paper for the plan went before Parliament it stated the money could be repaid as early as 1962. Included in the plan were the replacement of steam locomotives with diesel, large scale re-signalling projects, new marshalling yards for freight traffic, overhead electrification of trunk routes from London to Crewe via Birmingham and onto Liverpool and Manchester, the same for the Glasgow and East London suburban lines, as well as third rail electrification of Kent's main lines. In addition the plan included the construction of hundreds of new diesel railcars for provincial and non-electrified local routes and new freight wagons and locomotive hauled coaching stock. The final bill for all of this was a whopping one billion pounds. Whilst the plan had no direct effect on the Tyneside electric network it is important to understand the scale of the financial matters surrounding Britain's railways so as to understand the thought processes that would shape what was to happen next in the region.

The more observant travellers on the region had noticed one small change in 1954 when the British Rail built units appeared. A change to BR's numbering policy meant that vehicles built prior to nationalisation had a letter added to their number that identified their region of origin. In the case of the

By the time this was taken many of the surviving services north of the river were formed of just a single two-car set. This was the case when this train called at Backworth. (Bill Watson)

The 3:05pm Newcastle circular via the north side of the loop enters Manors formed of just a single two-car LNER unit. 21 May 1962 (Michael Mensing)

LNER-built units this was an E for Eastern region, so all car numbers on that fleet began and ended with an E. In June 1956 British Railways reclassified third class as second class; the move was a logical one as there had not been any second class on the whole of the railway network for several years and the provision of only first or third made no real sense.

First class was withdrawn from all North Tyneside electric services on 4 May 1959; however, the seats on the LNER units were not changed so you could enjoy sitting in a first-class seat for no extra charge. The South Tyneside service continued to advertise a first-class option and the solitary compartments in the British Railways units operating that line were not altered. This does raise the interesting question as to what happened when an articulated LNER unit worked the South Shields line, as they were reported as doing from time to time throughout their lives: if there was no longer any first-class accommodation on these units how would a first-class ticket holder get a refund? This change coincided with other changes to the timetable. The Riverside branch service was trimmed to a skeleton service, with only five trains from Newcastle to Tynemouth via the loop on a weekday, and an additional evening train ran on Tuesdays and Thursdays, whilst there were only three trains on a Saturday and the last one left Newcastle Central at 12 noon. In the other direction there were five trains but there was no additional one on days starting with T. Saturday witnessed only a single service, starting from Long Benton and making the journey around the coast to traverse the Riverside branch and terminate at St Peters at 7:18 in the morning. Unsurprisingly there were no services on a Sunday. The long-time regular on the Riverside line, the Controlled Set, was taken out of use and stored at Ponteland during 1959 but it was not officially withdrawn until February 1961, after which it went away for scrap. Whilst that set was in store the Riverside branch lost another station when St Anthony's was closed on 12 September 1960.

Tragedy once again struck the network on 10 January 1959 when a service entering Newcastle Central collided head on with an empty train that was being shunted. The front end of the incoming train bore the brunt of the collision and the driver was killed.

From 1960 the Tyneside units began to appear in a new shade of green as they became due for a repaint. This shade was a darker olive than the previous green applied to the fleet. Shortly after, small yellow panels on the lower cab fronts started appearing from 1962; these were applied to units both north and south of the river, but new paint could not hide the much bigger problems that had developed within British Railways. By 1961 the modernisation dream had turned into a publicly subsidised nightmare. The capital cost of the modernisation plan had failed to be repaid and passenger numbers had continued to dwindle. In 1960 alone the trading loss of the railways was £42 million. The Conservative Government of the time under Prime Minister Harold

June 1967 at Whitley bay in the last days of the Tyneside third rail electrics and the motorman is engaged with a young admirer of his train. (Colour Rail)

On 12 February 1967 E29127E leads a four-car formation into South Gosforth bound for Newcastle Central. (Colour Rail)

Macmillan had simply had enough of this situation and resolved to make sweeping changes in an effort to stop the rot. British Railways had undertaken a programme of route closures under the auspices of its Branch Line Committee since 1949 and up to 1962 this body had been responsible for instigating the closure of over 3,000 miles of rural lines that were either little used or duplicated other routes. The Tyneside electric network had simply not been considered as either of these so was perfectly safe, but higher management of all regions knew that not only did savings have to be made but that funding for improvement and upgrade projects would be tightly controlled. The modernisation plan was turning out to be an expensive failure; many of the diesel locomotive classes ordered under the pilot scheme had proven unsuitable and the replacement of steam was still a long way from completion. Freight traffic volumes had also taken a beating as many smaller loads switched to road as the plan had favoured larger block traffic and not the wagonload business that had historically been the railway company's bread and butter. In June 1961 Transport Minister Ernest Marples appointed Dr Richard Beeching as the new Chairman of both the British Transport Commission and the British Railways Board (BRB). The former organisation was going to be dissolved and the running of the transport system was to be devolved to dedicated boards: the BRB would be solely responsible for running British Railways. Beeching was no railwayman; he came from ICI and had a background in both engineering and physics, having obtained both his degree and PhD in the latter at the Imperial College. During the Second World War he had served on the Armament Design Section and joined ICI shortly after the end of hostilities. Beeching was known to have a shrewd head for business and Marples had clearly brought him in to shake up the running of the railway, which by then was operating at a loss of around £300,000 per day. Beeching told everyone who interviewed him that he believed the railways should 'Pay Their Way'. He set about his job of looking into the current state of the railway business and what he intended to do to turn it about.

On Tyneside, long before anyone heard the name 'Dr Beeching' there were those who looked into the crystal ball and knew change was going to have to be made before it was forced upon them. The commuter traffic on Tyneside remained healthy if not quite as buoyant as previous years. The number of journeys being made daily had decreased significantly from the heydays of the NER and LNER but was now at a plateau, mainly because the railways still had a vital role to play in ferrying commuters to and from work. Cars and buses could only steal so much traffic and had limitations of their own. Management knew that on the horizon was the question of rolling stock replacement north of the river as the LNER built units would be thirty years old in 1967; not only that but the substations and much of the conductor rail both north and south of the river would need renewal. Despite all the money spent on modernisation elsewhere during the previous six years, much of the electric network still retained an air of its Victorian and Edwardian ancestry. Most of the stations on the South Shields line continued to be lit by gas and a few still had their LNER name boards. North of the river many of the stations were in a neglected state. The Tyneside electric network was going to need a lot of investment within five years to keep going and now was not a good time to consider asking for the funds to do this. In December 1958 an experiment was conducted whereby a small selection of electric services were replaced by diesel multiple units (DMU) for two weeks. This would allow for a full evaluation on the suitability of these units to not only keep time on these services but also to compare the associated running costs between diesel and electric traction. Seven weekday services and four Saturday trains were replaced as part of the trial. DMUs also replaced all of the South Shields services between 22 February and 3 May 1959 but this was more for logistical reasons as the High Level Bridge was closed for maintenance. The service had to be diverted from Newcastle Central over the river via the un-electrified King Edward bridge, where they reversed at Gateshead West to get onto the South Shields route. The results of these trials were never announced but in September 1960 a further trial was conducted with a three-car DMU running in place of an EMU on certain North Tyneside services on the 7th and 8th and then on South Tyneside services on the 9th. Again no information was forthcoming on the outcome of these trials. Changes were made in the 1961/62 winter timetable when the South Shields line went from a twenty minute interval service on a weekday to a roughly half hourly one, this meant fewer units were required for traffic, and a longer dwell time at South Shields was allowed before returning to Newcastle, On Sundays Felling station was closed and trains were no longer required to stop there. North of the river the last of the Pram Vans converted from the fire replacement stock were withdrawn from service: cars 29376 and 29387 were withdrawn in July, the other cars, 29388 and 29390 had been withdrawn during May 1960.

December 1962 was when the announcement came that the electric services were going to be progressively withdrawn on Tyneside. At first the public misunderstood the announcement to mean the lines were going to be completely closed, or perhaps it was that British Railways had not communicated their intentions clearly. Either way there was much uproar and protest meetings were held, possibly even attended by those who no longer used the railways and were now concerned about the extra cars and buses that would clog up the streets. British Railways clarified its position by affirming the network was to remain open to passengers but that the services would be operated by diesel railcars instead of the electric units currently in use. The reasons given were that the cost of replacing the old stock could no longer be justified when there were a surplus of diesel railcars on hand that had

June 1962 at South Shields and the BR units have just over six months' service left on Tyneside. (Colour Rail)

been displaced from routes that had closed. In addition the cost of supplying the electric current to the line was some £86,000 per year whilst fuel oil for the railcars would amount to only £14,000 for the same period.

Before going on it is worth taking time to reflect, with the benefit of hindsight, on this decision. Many local people, enthusiasts and historians have called this decision both wrong and short sighted. Electric traction was meant to be the way forward for the suburban railways; the NER knew that as did many other companies, which is why the network was electrified. Surely the railways should be

The 2:22pm from South Shields crosses the High Level Bridge on the final leg of its journey to Newcastle Central, on 28 May 1962. (Michael Mensing)

The End • 95

The eastern end of Newcastle Central station on the evening of 27 June 1960. The DMU on the right is bound for Sunderland whilst a north Tyneside electric service arrives on the left. (Ron Fisher)

Now sporting a small yellow warning panel a pair of articulated units enter Manors station bound for Newcastle. I'm not sure you would get away with running with the doors open these days. (SERA Archive)

investing in this network to bring it up to date instead of replacing it? These arguments are perfectly valid but to counter them it is important that first we put ourselves in the shoes of those running the railway in the early 1960s. There are a few hard to swallow but undeniable facts that cannot be ignored. The first is that the financial case for replacing the LNER stock could not be justified. Given that the new units would have needed to be in place by 1967/68 this would mean the BR workshops building a fleet of between fifty or sixty four-car units. Following the previous policy of equipping the network with standard BR builds this would have meant a fleet of units of either the Southern Region CIG or VEP fleet that were either in or entering production at the time. On the other hand if you had a fleet of perfectly useable diesel railcars that were available straight away would you seriously spend money that you did not have to build the new trains? It simply did not make economic sense. Then one also has to consider the investment needed to keep the network electric. The railways had to justify all new spending and there had to be evidence that the money would be recouped by ticket sales. The Tyneside network was essentially a Victorian railway and it could not deliver people to the door of their destination in the way the car and bus could. Indeed both of those forms of transport had eliminated the electric trams on Tyneside before the Second World War started. For the railway to become more convenient it would have to have more stops and better access to the city centre. Given that on this Victorian railway the electric services shared the lines with both freight and other passenger services there was not the scope to remodel the lines in the early 1960s to meet these needs, nor was there the cash or the desire to do it either. As hard as it may be to accept, the NER's electrified network had served the locals well but its time had come and something new would have to replace it. In the meantime the network would have to endure a few years of stagnation whilst the powers that be decided what that something new would look like.

The winter of December 1962 going into January 1963 was one of the worst for many decades. Across the country temperatures plummeted, lakes and rivers began to freeze over and then the snow fell. It was against this backdrop that the last third rail electric services ran on the Newcastle to South Shields line, giving many photographers a chance of the 'Christmas Card' type view as they recorded the passing of this era of the region's rail history. On Sunday 6 January 1963 the last electric service operated by British Railways crossed the River Tyne, and immediately after that day the line was wholly the preserve of DMUs. Parcel services went over to steam or diesel operation. No time was wasted dismantling the apparatus of electrification and by the end of the following week the conductor rails had been removed from over the High Level Bridge; eventually they were removed from the whole length of the line. The fifteen two-car units built in 1954/5 for the route were initially stored at South Gosforth along with the Motor Luggage Van E68000; the two-car units had seen less than eight years' service so were suitable to rehabilitate on the Southern Region where their close cousins were still hard at work.

In a few weeks all fifteen had been made ready for haulage south and delivered to Eastleigh Works, where they would undergo a programme of overhaul and de-Tynesiding to make them as similar to the rest of the EPB fleet as possible. The first-class compartment was declassified, the route and headcode indicators on the cab fronts were removed, replaced with a two digit route blind that was two-thirds the height of the one fitted to the original two-car sets built for service in the south. The longer than usual brake van could not be altered so there were features that would ensure these units retained elements of their early life for the rest of their careers and were always noticeable to those who knew their history. The rebuilding work carried on for nearly eighteen months until the last of the fifteen sets was released from Eastleigh in August 1964. Unlike on Tyneside they were allocated unit numbers on the Southern Region and were classified, as the other units purposely built for the Southern Region were, as 2-EPB and allocated unit numbers 5781–5795. They got to see out their thirty-year expected life span in passenger use, being withdrawn from service in 1985. One vehicle, a Driving Trailer Second, was given a further facelift and continued in passenger traffic as part of a different unit, numbered 6409 until 1993. The rest saw additional use in departmental traffic as tractor and test units with one pair of Driving Motor Brake coaches being reformed back to back as a sandite and de-icing unit. Numbered 017 it worked out of both Bournemouth and Wimbledon depots, being renumbered to unit 930 201 in 1999 until finally withdrawn in 2006. It was scrapped at Immingham that year. One test unit was formed of the Driving Trailer Composite of 5793 and the Driving Motor Brake of 5791 and it was this that the author was instrumental in saving for preservation (see later chapter).

The Eastleigh built Motor Luggage Van was not suitable for use on the Southern Region as it was a very different design to the fleet of ten vans used in Kent for boat train and parcels traffic. Those vans had only one power bogie and were fitted with traction batteries to enable them to work on the non-electrified quay line at Dover. E68000 did not have this feature so would not have been able to integrate itself into that fleet. It was decided to send it to the London Midland Region's third rail lines on Merseyside where it could handle additional parcels traffic between Southport and Liverpool. The van was modified by having the destination lights and route blinds removed and marker lights fitted on the front above each buffer. In the event it proved too non-standard for that region; there was insufficient traffic to justify its retention and it was withdrawn from service in 1966, being sent to Willoughby's scrap yard in Chopington, back in the North East, where it was broken up in late 1966.

Some of the North Tyneside electrics were to carry on for a few more years. The same changes that did away with the South Tyneside electrics reduced the services north of the river too. Eleven Type C units and four Type D units were withdrawn in 1963 and sent away to be scrapped at Blyth. These units were targeted as they were the least operationally flexible, having driving cabs at one end only. The service was trimmed to reflect the reduction in stock, with many off peak trains on the coastal circle formed of just a single two-car set, although longer trains would still appear at busier periods.

On 27 March 1963 Dr Richard Beeching published his report entitled *The Reshaping of British Railways*. In this he proposed the closure of 6,000 miles of mainly rural and secondary lines, which included nearly 2,500 stations.

Among those lines listed to face the Beeching Axe was the Riverside branch, which had suffered a reduction in service a few years earlier. The main coastal circle was not intended for closure, as perhaps the unilateral actions to dieselise the routes had kept them off the list, something we shall never know for sure. After much deliberation and campaigning it was decided to reprieve the passenger service, as much as there was of it, on the branch and so the North Tyneside electric system would remain intact for the last few years of operation.

Despite the end looming for the LNER-built units, a handful had their whistles removed and replaced by two-tone warning horns as fitted to diesel locos and

A pair of North Tyneside LNER two-car articulated EMUs, with E29160E leading, arrive at Heaton on a stopping service to Newcastle Central, c.1964. (Rail Archive Stephenson)

The 11:45 ex-Newcastle circular is on the inward leg of its journey and arriving at Manors station on 28 May 1962. (Michael Mensing)

units of the time. The horns were mounted on a bracket below the non-driving side of the buffer beam and operated by a two way air valve in the driving cab. One other modification was to change the E prefix on the vehicle numbers to NE denoting they were North Eastern region vehicles. Quite why this was done was a mystery as the North Eastern Region had been absorbed into the Eastern Region of British Rail by then and in the end only three units got the new prefix applied. In March 1966 with the winter over, the two ex-NER motor luggage vans that had been converted to de-icing vans were withdrawn from traffic after a career of sixty-two years. One of them, 900730, formerly NER 3267, would be saved for the nation and after a period in store at Gosforth, Heaton and Monkwearmouth would eventually be restored to North Eastern Railway red and cream. It is now proudly on display

The last winter for the LNER stock. Pictured in early 1967 a unit sporting a small yellow panel enters Newcastle Central. (Author's Collection)

at the Stephenson Railway Museum at Percy Main; it has none of its original electrical equipment or cab controls as they were long gone as part of its conversion to a de-icing van, but the exterior of the vehicle is pretty much as it was when built and it is a unique survivor of the early days of electric traction in Britain. And so the remainder of the fleet entered their final year in traffic. An extra five minutes was added to the running times for the stopping trains so the DMUs could cope with the schedule; more services were being placed in the hands of the diesel units, with more electric units being withdrawn in the spring of 1967.

The final day came on Saturday 17 June 1967. In the morning an unusual working took place when a local parcels service from Newcastle to Durham and back, far away from third rail territory, took Class 17 'Clayton' diesel electric with LNER-built Motor Luggage Van E29467E as its trailing load, reportedly due to a lack of suitable loco-hauled stock. The van was parked up at Newcastle Central after its trip and withdrawn the next day after working back to South Gosforth. The final passenger working was the 18:15 departure from Newcastle Central, out via Wallsend to

The classic view taken from the Castle Keep of a LNER unit departing with a North Tyneside service. The snow on the ground puts this as the winter of 1967 so the electric services were only a few months away from withdrawal. (Author's Collection)

This was the view from the Castle Keep looking in the other direction away from Central station. In this image dated May 1964 a pair of North Tyneside units share the viaduct with an A3 Pacific. (Ron Fisher)

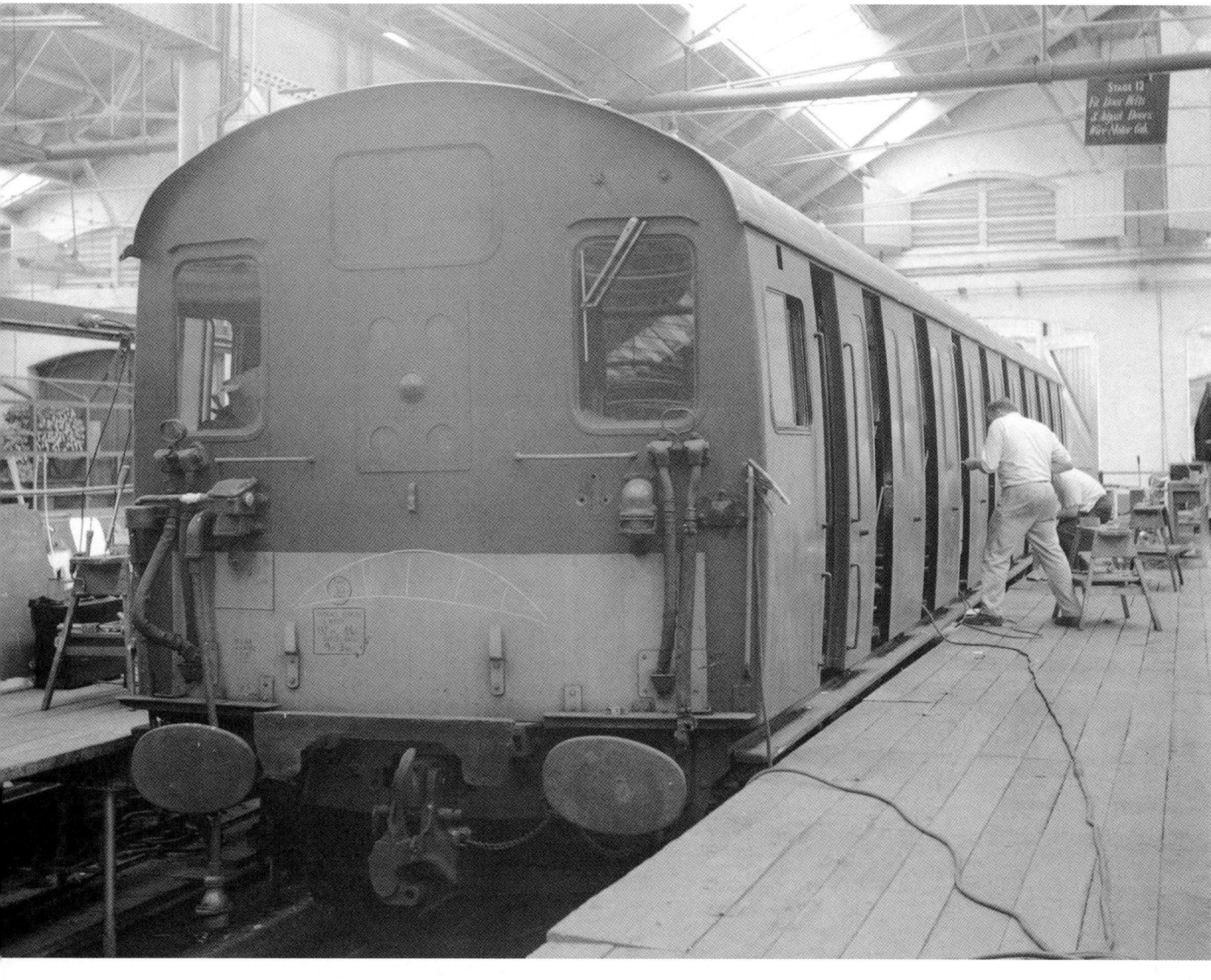

A rare view of Driving Trailer E77104 undergoing conversion for use on the SR at Eastleigh Works, 7 August 1963. (Robert Carroll Collection)

Tynemouth and then back via the north side of the loop to Benton and back to Central. Formed of an eight-car set it was crammed with both the regular users of the services and dozens of enthusiasts who had come out to travel on the last train and see the real end of an era. The train arrived back at its destination. The passengers disembarked, photographs were taken as keepsakes of the occasion by those who cared and then the stock ran empty to South Gosforth car sheds. It was all over.

Over the next few months units were stored at Heaton, Gosforth and Ponteland whilst their disposal was arranged. It was a busy time for the scrap merchants, with hundreds of steam locos, wagons and carriages coming out of traffic, as well as a few of the least successful designs of diesels from the Modernisation Plan. The bulk of the LNER units went to Blyth to be scrapped but a handful made the journey north of the border to be broken up in Scotland. One last memento appeared at Manors station waiting room, when a set of bucket seats recovered from a redundant unit was installed there briefly.

This former South Tyneside luggage van moved to the Liverpool area and was employed on the Liverpool to Southport route for a short time after being displaced by the de-electrification of the route it was built for. Here it is seen near Birkdale on 8 May 1965. The modifications made to it are evident but it was to prove a short-lived relocation. (Jim Peden)

Going for scrap! Here we see 8X67 0500 Heaton to Coatbridge entering the loop at Brampton Fell conveying six twin articulated EMUs and two Parcel Vans. The train loco is Class 45 No. D83 and the van is E29168E, and as can be seen the Parcels van has suffered from the local vandals. (Keith Long)

A withdrawn two-car articulated set is pictured dumped at Heaton in the company of several withdrawn steam locomotives. (Michelle - 70023venus2009)

A former South Tyneside EPB unit, 5794, leads this four-car formation on a Waterloo to Chessington South service between Tolworth and Chessington North.
(Brian Morrison)

A Waterloo to Guildford via Cobham service pulls away from Clapham Junction on a winter's day. The leading unit is 5787, one of the former South Tyneside 2-EPBs.
(Brian Morrison)

Sandite and de-icer unit 930 201, formerly 930 017, stands in Clapham Yard c.2002. This unit was formed of a pair of former South Tyneside EPB Driving Motor Brakes and was the last ex-North Eastern third rail stock ever to operate on the national network. (Graeme Gleaves)

An era quietly comes to an end as 930 201, the last of the Tyneside electric units in operation, was quietly broken up at Immingham in company of southern built EPB stock when Network Rail disposed of its heritage sandite and de-icing fleet in 2006. Pictured here awaiting the final torch. (Graeme Gleaves)

Chapter 12
THE LAST LAUGH

British Railways wasted no time in removing the evidence of electrification on Tyneside. The conductor rails were lifted and disposed off for scrap along with all the heavy duty cables that fed them. The porcelain insulators the live rail sat on were ripped up too and left their scars on the sleeper ends where they had been secured with steel screws. In a few isolated locations the odd discarded section of conductor rail or its supporting porcelain insulator pots could be noted lying in the centre or by the side of the track for a few months, where they had been overlooked in the dismantling process. The substations would also be stripped of their contents and demolished. The train services, now predominantly in the hands of Metro-Cammell built diesel units instead of that builder's electric sets, continued to operate a service that was a shadow of the one provided in the heyday of electrification. There was no shortage of the diesel units as a result of the mass closures that had come about from the Beeching report and it was not uncommon for units operating on Tyneside to still be displaying route maps of lines in Devon and Cornwall inside. The diesels operated at a much slower pace than the electric trains they replaced. They were not designed for the quick acceleration of point to point suburban working, and no diesel unit ever could be, so the timetable was padded out to suit their limited performance. British Rail launched a marketing campaign to promote the lines in October 1970 when it gave the people of Tyneside 'The Tynerider'. This was a serious and moderately successful attempt at rebranding the services to give them a local feel. Orange leaflets and timetables were distributed and the orange themed branding was extended to the suburban platforms at Central station and on the trains themselves, despite the inevitability that many units spent time on diagrams that were over routes not part of the Tynerider scheme. The frequency of trains was restored to twenty-minute intervals between 06:00 and midnight on the north side of the river, whilst the South Shields line remained half hourly. The emphasis was on promoting travel throughout the day and night and not just during the busy peak periods. Trips for work, shopping and nights out were catered for and an overnight service of three trains on the coastal loop were run on six days of the week for those who stayed out very late, or got up very early. The result was that passenger numbers actually began to go up, rising to nearly eight and a half million journeys in 1971 compared to 7.8 million in the previous year.

The Riverside branch had survived the initial recommendations of the Beeching report and had stayed open to passengers with its skeleton service continuing to take workers to the river front-based industries on the north bank of the Tyne, but as the 1960s ended and the 1970s began

The advertising for the Tynerider DMU service that replaced the electric units. (British Rail)

the line once again came under scrutiny. The line still had some freight traffic but the passenger service had been trimmed to two trains each way Monday to Friday to coincide with shift times at the shipyards that were by then the only source of passengers. The stations had no residential areas in their catchment zones and the proximity of the better-served lines on the coastal circle meant passengers would use those instead. Subsidy for the line was withdrawn by the newly formed Tyneside Passenger Transport Executive in 1972 and a new road that could offer alternative transport to the areas served by the line opened the same year. The Riverside branch had been on borrowed time for nearly a decade but time ran out for it on 23 July 1973 when all passenger services were withdrawn. Freight would continue for another fourteen years over parts of the route and it would take the best part of a decade to demolish the stations.

During the first five years of diesel operation wheels were moving within both local and national government. The first significant step came when the Labour Government of the time passed the Transport Act in 1968. This created several regional bodies around the larger conurbations to co-ordinate the running of local transport. In the case of the North East the relevant body was called the Tyneside Passenger Transport Executive (TPTE) and they oversaw, but did not directly run, the local train services as well as taking over the running of local municipal buses and ferries in the area. Effectively they were responsible for the area's local transport network including how the services interacted with each other and other services such as roads. The TPTE came into being on 1 January 1969. They began a much needed review of public transport in the area and found that the roads in the city were congested; some of the factors driving this were the lack of river crossings, road improvement projects that had not been delivered and over reliance on buses due to inadequate rail services serving the city centre. Options for improving the rail routes were considered and ranged from converting some routes to dedicated bus ways to the provision of a new light rapid transit rail network based upon the former electrified suburban lines. It was this latter suggestion that won the day.

The proposal that went before Parliament was a plan for the new operation taking over 41km (25 miles) of existing BR lines and 12.8km (8 miles) of newly constructed line, most of which was going to be in tunnels under the city centre. It was this new section that would give the network the edge its predecessor never had in being able to bring passengers directly into the centre of Newcastle. Also included in the scheme was a new bridge over the river linking Newcastle and Gateshead, as the High Level Bridge was still needed as a dedicated route for mainline British Rail services. A grant was obtained by the Tyneside PTE for the bulk of the construction costs, which were estimated at £65 million, and the Tyneside Metropolitan Railway Act was passed by Parliament, receiving Royal Assent in July 1973. Once again the North East was breaking new ground as nothing like this had been attempted in the UK before.

The local government re-organisation of 1974 saw the Tyneside PTE have its area of scope extended to include Sunderland so the organisation was renamed the Tyne and Wear PTE. The railway they were building was christened after the expanded authority and became known as the Tyne and Wear Metro, or simply known locally as 'The Metro'. The Metro was to take over both the coastal loop line north of the river and the South Shields line. Both would require substantial rebuilding of not just the permanent way but the stations too. The Metro was to be electrified, not with conductor rails but with overhead wires, energised at 1,500v DC. The return of conductor rails had been considered but ruled out on a number of reasons: reduced energy efficiency, the vulnerability to ice of conductor rails, the cost of extra substations and finally on grounds of safety. As the network was no longer electrified by this method only an extension to an existing third rail network would be authorised by the then Chief Inspector of Railways and the Tyneside network did not fall under this scope. Overhead wires were also safer where the railway crosses roads and within depots and sidings. The higher voltage available to an overhead wire system reduced the number of substations required to just eight. The bulk supply point for the central part of the network would be at South Gosforth, where the supply is delivered by the North Eastern Electricity Board (NEEB) at 33,000v AC which is then fed to the substations via 11,000v

AC ring cable feeders which the substation converts to two supplies: one is the 1,500v DC traction supply whilst the other is a 415v AC supply for lineside signalling, lighting in the tunnel section and other control equipment for the infrastructure. In the outer area the substations are fed directly by the NEEB and their locations were carefully chosen to ensure this could be achieved based upon existing local power supply infrastructure.

The work required to convert the former BR lines was considerable and involved far more than simply erecting overhead wire catenary equipment. In preparation for the works the lines were closed in sections and replacement buses provided ahead of the Metro opening. The process began on 23 January 1978 when the direct line between Newcastle Central and West Monkseaton was closed, leaving the southern and coastal sections of the coastal loop as an out and back service from Central to West Monkseaton via Wallsend, but this was cut back to terminate at Tynemouth from 10 September 1979. This arrangement continued until 11 August 1980 when the first section of the Metro opened from Monument station via South Gosforth to Tynemouth. On the same day the southern section of the coastal circle from Tynemouth to Newcastle Central was closed. That line re-opened on 14 November 1982 all the way through the city centre to a new terminus at St James, adjacent to the Newcastle United stadium. Between these two sections opening there was a new line added to the list of electrified lines on Tyneside. The branch from Newcastle to Ponteland had lost its passenger service as early as 1929 but the line remained in use up until the late 1960s for freight traffic. The Metro had ambitions to eventually reach Newcastle Airport using the alignment of the Ponteland branch for most of the route and it opened the first section of this on 10 May 1981 as far as Bank Foot. The line diverged from the main metro route just to the north of South

The last day of DMU operation on the South Shields line was 30 May 1981. Class 105 cars E56429 and E51282 are seen arriving at South Shields with a service from Newcastle; these units were not common on the Tyneside routes. (Graeme Phillips Collection)

Platforms 1–4 at Newcastle Central pictured on 6 October 1982. All services from these platforms have ceased and the site will be redeveloped a few years after this image to become a car park. (Graeme Phillips Collection)

Gosforth and then headed north west with intermediate stations at Regent Centre, Wansbeck Road and Fawdon. As this line had been out of use for some time the Metro conversion work could be carried out without the need to plan around alternative services for passengers. A few months later on 15 November 1981 Monument station was opened giving the network a convenient station for the city centre for the first time ever.

The tunnel sections were by far the largest engineering challenge. These had to be excavated with minimal disruption to the daily life going on above in Newcastle and were bored through the clay beneath the city and lined with a mixture of steel or concrete lining sections. The tunnels made the former suburban platforms at Newcastle Central redundant and these were eventually lifted and filled in. The stations at St James, Central Station, Manors, Monument, Haymarket and Jesmond were all new construction and owed nothing to the previous alignment or stations they shared names with from British Railways days. All were subsurface with escalators and lifts linking them to street level. The stations were brightly lit with the Metro house colours of yellow and black used for signage, including the station name boards. Like all stations and trains on the network they were completely non smoking areas, the Metro beating the London Underground to be the first smoke-free railway network. A station name that re-appeared on the route map from days gone by was Byker; the original had been located on the Riverside branch but the new

Tunnelling work in progress under Newcastle to construct the Metro. (Newcastle Central Libraries)

station built for the Metro was in a different location. The line from Newcastle emerged from the tunnel section to the east of Manors and followed a new alignment as it made for the coast. This involved one of the other great engineering challenges in the shape of the Byker Viaduct that took the new formation across the Ouseburn Valley. The viaduct is 800 metres (2,625 feet) long and at its highest point is 70 metres (230 feet) above ground. The structure was another first for the Metro as it was made using cantilevered concrete sections with glued joints, the first time such a method had been used in the UK. A total of 253 concrete sections were cast on site and lifted into place, secured using a combination of epoxy resin and steel cables. Byker is the first open air station in this direction and was followed on the route by another new station called Chillingham Road. Both these new stations replaced the former one at Heaton which did not get rebuilt for the Metro and never re-opened.

By now the old NER track bed was being used and the next stations were Walker Gate and Wallsend, as they had been when the line was last electrified. After Wallsend a new station was provided called Hadrian Road, before two stations familiar from a previous life at Howdon-on-Tyne and Percy Main; both of these stations were completely rebuilt from their BR state and have none of their original features. Between Percy Main and North Shields a new station at Smith's Park was provided; this served the Meadow Well housing estate and the station would eventually be renamed Meadow Well in October 1994.

Publicity still of Metrocar 4002 on the test track. (Tyne & Wear PTE)

Interior view of the tunnel section built on the Metro test track to evaluate noise and train behaviour. (Newcastle Central Libraries)

North Shields station was another one completely rebuilt for Metro traffic save for some of the 1960s' era platform canopy. The entrance hall and platform structure was all new construction. The next five stations were exactly the same as they had been at the end of the previous electrification: Tynemouth, Cullercoats, Whitley Bay, Monkseaton and West Monkseaton. All the stations retained a degree of their original structures, some more than others.

Whitley Bay, Tynemouth and Monkseaton were the least molested whist the LNER station entrance at West Monkseaton was retained but brand new Metro platforms were provided. Shiremoor was a new station provided for the opening of the Metro and situated half a mile east of the former NER station at Backworth which it replaced; the new location was more convenient for the residential areas. The next station was Benton, another name from the previous incarnation of the route, unlike the next station at Four Lane Ends, which was opened to provide an interchange with local bus routes. Longbenton had been opened by the LNER and was retained by the Metro more or less intact; the platforms were truncated for the shorter Metrocar formations but the rest of the structure was re-used. This was not the case at South Gosforth where the whole of the station was obliterated and replaced with one to Metro design. Oddly the original NER footbridge to link the two platforms was re-used but nothing else of the original station survived. Between South Gosforth and West Jesmond a new station was built at Ilford Road. West Jesmond was the last station on the original NER formation before the line entered the new tunnels for the journey to Newcastle. It retained its original buildings with new Metro-style platforms provided. As mentioned, the platforms that were retained from BR days were truncated to a length of two Metrocar sets, the longest train the network would operate, and a far cry from the days of eight-coach suburban trains under the LNER. The new-build stations and those former BR stations that had their platforms demolished had a new structure installed that was long enough for the new trains and no more. Some of the old BR stations simply had the excess platform screened off and left *in situ* both unused and un-tended, Tynemouth and Whitley Bay being two such examples of this. Cameras and TV monitors or large mirrors were provided as the Metro was operated with a driver only and this equipment enabled the driver to check passengers had completed boarding or alighting before the train doors were closed. Other facilities were the provision of ticket machines as the Metro had no staffed ticket offices and used a simplified fare structure compared to BR and which split the network into three zones. A passenger bought a ticket that covered the zones they were travelling from and to so the choice of fare went only as far as single or return for the relevant zones.

The South Shields line presented its own set of problems for the construction team. A new bridge was required to cross the River Tyne. Designed by W. A. Fairhurst and Partners the bridge was built of predominantly steel

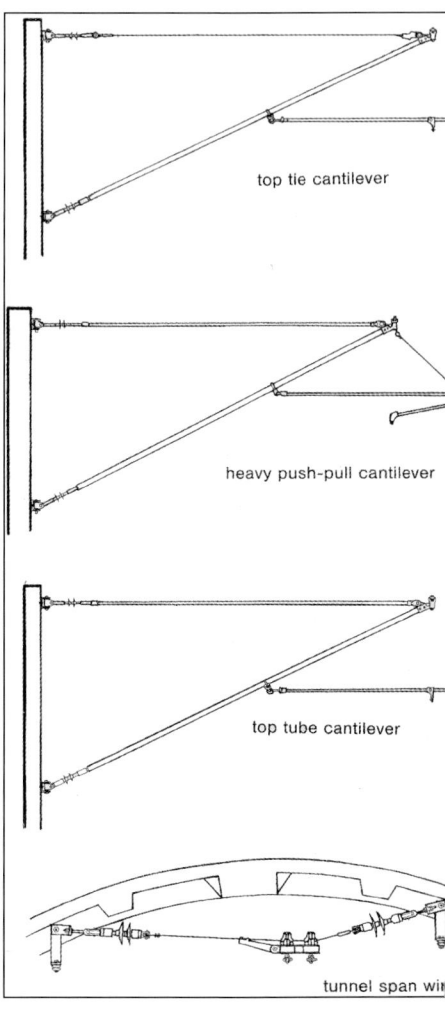

Diagram of the four different types of overhead line support structure used on the Metro. (Author's Collection)

with pre fabricated box sections being supported on two concrete pillars and steel truss sections. Construction began in 1976 and was on both banks of the river simultaneously, with the aim of the two teams to meet in the middle. This was achieved in August 1978 and the structure was equipped for the operation of Metro trains and opened with the first section of the former South Tyneside route

Aerial view of the South Gosforth car sheds, ordered by the NER, opened by the LNER and now the home for the Metrocar fleet. (Tyne & Wear PTE)

on 15 November 1981. The official opening of the bridge had been carried out just over a week earlier on 6 November by Queen Elizabeth II, whom the bridge is named after. The bridge cost a total of £4.9 million to build and provided the only section of daylight between the two tunnel sections either side of the river. The old BR route through Gateshead could not be used as it was required for exclusive use of BR trains so the route had to be excavated in tunnels. This involved the filling in of excavations and evidence of mining, some going back to the middle ages, in the rock forming the south bank of the river. The provision of a new sub-surface Gateshead station enabled it to be far better sited for the main shopping centre than the previous one and excellent provision was made for a local bus interchange. The tunnel then continued to just before the next new station at Gateshead Stadium, after which the line took up almost its former BR formation as it had to run parallel to the BR lines which were still in us as far as Pelaw.

The stations at Felling and Heworth were retained, albeit in rebuilt state for Metro operations. Pelaw was not included as a station for the Metro so it was at Heworth that the first stage of opening of the South Tyneside Metro terminated. The next phase onto South Shields would not open until 24 March 1984 as a significant amount of the route had to be built from scratch to overcome the shortcomings of the old BR line. The next stage had stations at Hebburn and Jarrow as before but completely rebuilt on their relative sites. A new station was provided at Bede where a trading estate was located. Tyne Dock station was also retained on its original site but again, was completely rebuilt. The line beyond Tyne

Ticket offices and ticket barriers on the Metro were automated from the start. (Tyne & Wear PTE)

The control centre that regulates movements in, out, and around South Gosforth depot. (Tyne & Wear PTE)

Dock follows a different alignment to the route of the LNER South Tyneside electrification and was partly built on the track bed of the Stanhope and Tyne route. A new station was built at Chichester with facilities for bus interchange following the theme of integrated local transport. The new formation then continued to a single platform terminus at South Shields, located above King Street where the main shopping area is located. This station would subsequently close and a new Metro station would be built to the west of the previous one on Keppel Street, where there is now a combined Metro and bus interchange. The new station opened on 4 August 2019 and is the sixth station since 1835 to bear the name South Shields.

As mentioned Pelaw nearly didn't get a Metro station. This was in part due to the spiralling costs of the whole Metro project. The original estimate for both construction and equipping the system of £65 million proved woefully inadequate. In truth the project suffered from increasing costs brought about by high inflation and longer than anticipated build times. There was a real danger that the whole project could be either cancelled or truncated to save expenditure. Common sense did prevail and the network opened in the stages described and has become both a success and as much a part of life on Tyneside as the NER's electrified lines did. Pelaw did eventually get its station, a brand new construction as the BR station had been demolished. The Metro station opened on more or less the same site on 16 September 1985.

Chapter 13
METROCARS AND BEYOND

The Metro needed a new type of train to operate the dedicated services. It would have to be designed as a bespoke item as the system as a whole was unique, and no existing off the shelf rolling stock design could be adapted. From the outset Metro trains would need to be one person operated as they did not have guards, only drivers. They would operate in trains of either two- or four-coach formations powered by overhead wires and had to be capable of both rapid acceleration and braking to meet the demands of the start-stop nature of the timetable. Trains would get busy at peak periods so provision for standing room under crush loading had to be a consideration, with sufficient seating to meet the requirements of the off peak traffic. Whilst there was no real direct equivalent for this type of train in the UK, other than possibly the London Underground, there were certain parallels in Europe, and the Metro aimed to use the best practice along with tried and tested technology when considering the design of the Metrocars. Tenders for the construction of the rolling stock were put out and the preferred bidder was a company well known to students of Tyneside's electric trains: Metropolitan-Cammell of Washwood Heath near Birmingham. They may have relocated from the Black Country but this was the same Met-Cam that had built the articulated stock for the LNER, and now they were going to build articulated two-car units for the Metro.

Construction started with two prototype sets built in 1975 which were delivered to the Metro test track at Middle Engine Lane, just north of Percy Main. The line was formerly part of the Blyth and Tyne Railway that had been built for the transport of coal to the river. The line had been absorbed by the LNER and then by BR afterwards, but closed as the collieries it served had been run down. The test centre was equipped with overhead wires and a shed to house the two prototype units, and running from this was a test track of approximately a mile and a half that stretched towards Percy Main. This was electrified to the same standards as those planned for the Metro and the whole purpose of the facility was to rigorously test all the equipment the Metro would use in anger so any snagging issues could be resolved before the construction of the lines began in earnest. The test track featured a Metro-style station and even a short mock tunnel section.

In addition the centre acted as a training facility for the staff who would run the network day in and day out. The first set numbered 4001 arrived at Middle Engine Lane on 19 May 1975 where it was set up and statically tested. It made its first movement under its own power on 10 June just over a month before it was joined by the second set 4002, which arrived on 21 July 1975. The test centre was operational for five years, closing in July 1980 as construction work on the ex-BR depot at South Gosforth was advanced enough to allow the Metro to take possession during 1979. The two prototypes were moved by road to South Gosforth either side of Christmas 1980 with the last vehicle being delivered there on 8 January 1981. The test track found another use in time and is now the Stephenson Railway Museum, where a large collection of local railway items are gathered, ranging from 'Billy', one of the earliest locos built and used for hauling coal wagons, all the way to diesel shunters used as the station pilot at Central Station. One item of note is the ex-NER Motor Parcels Van which was saved for the nation after its role as a de-icing van came to an end in 1966. It is beautifully restored and on display wearing its original NER red and cream livery.

On 8 July 1978, Metrocar 4002 is on show at the Metro Test Centre, West Chirton. (Neil Sinclair)

The driving console of a Metrocar at West Chirton. The larger lever to the left of centre is the combined controller for brakes and power whilst the smaller lever to the right of it sets the direction of travel. The radio panel is on the right. (W. Welsh)

Metrocars and Beyond • 117

Prototype Metrocar 4001 on trials on the test track. Note the maroon underframe and running gear that was not adopted on the final version of the livery. (Brian Pearson)

When the test track was no longer required it became a heritage railway and the Stephenson Railway Museum. The southern end of the heritage line passes under the Metro here, just east of Percy Main station, and a pair of Metrocars in Woodpecker advertising livery are passing overhead. (Graeme Gleaves)

Metrocar 4001 at the former Blyth and Tyne railway station at Jesmond that dates from the line's opening in 1864. This station closed on 23 January 1977 when construction of the first stage of the Metro system began. It was replaced when the Metro re-opened in 1980 by an all new station a short distance to the north. The former station is now a restaurant called The Carriage. The track is still in place today and is used to move cars that are out of service back to Gosforth Car Sheds. (Russell Graydon)

Another view of 4001 at the former Jesmond station. The train was there in connection with an event at the adjacent council offices. (Russell Graydon)

As the 1970s drew to a close, construction of the production Metrocar units had begun at Washwood Heath. In addition to the two prototypes a total of eighty-eight sets were ordered. The original plan for 130 sets had to be trimmed to save costs, which were spiralling at the time. As mentioned they were two-car sets with the two cars articulated over a central bogie. Passengers could pass between the two vehicles as the interior was entirely open plan, except for the driving cabs that were a lockable closed compartment on the left-hand side of each cab front. This left the right-hand side free for passenger space so it would be possible to sit up front on a Metrocar and enjoy that view. The driver's compartment was fitted with opaque glass so passengers did not disturb, or even observe, the driver at work. The two-car set had seating for 84 passengers but there was standing room for over 100 more under crush loading conditions. The entire unit is 27.8m long. (It should be noted that all dimensions and distances on the Metro are measured in metric units so this chapter will do the same.) Of the three bogies the unit sits on, the centre one is unpowered, with both the driving end bogies being equipped with a single 248hp traction motor mounted longitudinally with drive to both axles of the relevant bogie achieved through right angle gear boxes. The two motors are more than capable of accelerating the unit up to its maximum speed of 80kph. The two motors are permanently coupled in series and acceleration is by means of camshaft control which steps out traction resistances as road speed increases. A similar system of motor control was first used on tube trains in London during the mid 1930s and the Metrocars employed this system as part of the remit for tried and tested technology. With the braking system the Metro cars have a mix of rheostat, which uses the motors of the train to slow it down, and an air friction brake which comes into play when speed is below 20kph. Each bogie is also fitted with a magnetic track brake that can stop a train in an emergency in a distance of 150m; obviously this brake is only to be used in an emergency. Each set is fitted with a single compressor to charge the air system to operate the brakes, along with the control pressure needed for electrical equipment switching, window wipers, warning horn, and to drive the pressure heating system. Construction of the frame and bodyshell is a mix of steel, alloy and aluminium. The units have no buffers and a BSI automatic coupler not only physically couples sets together but houses the connections for air and electrical lines between them. In passenger service only two sets are coupled together but up to three can be coupled for empty stock workings.

Inside the Metrocar extensive use is made of plastics for both ease of cleaning and ease of replacement. The lighting is fluorescent and public address equipment is fitted as standard. Each articulated half of a unit has four pairs of air-operated passenger doors (two pairs on each side) that open outward of the body side, rather than sliding doors that are contained within the bodyshell as were favoured on both the London Underground and British Railways for their suburban units. The Metrocars have pressure-operated heating and ventilation, as well as under-seat heating but the hoppers on the upper part of the large body side windows can be opened by passengers to increase airflow inside the car if so desired.

It is inside the driver's cab that the most modern advances have been made. The power and brake controllers are combined into one lever that applied the brakes when pushed forward and accelerates the train when pulled back. As the trains were to be worked by only a single crew member it was a requirement that a 'ship-to-shore' communication system was fitted. This is in the form of a two way UHF radio link, and is a follow on from the system developed by the Tyne and Wear PTE for use on its bus services. The system allows not only the driver and control centre to talk to each other but also for the control centre to make PA announcements on the train if required. The driver also has the option to call on maintenance staff via the radio in the event of the train having technical difficulties.

The Metro places great emphasis on passenger information screens at stations, and to provide running information about trains each Metrocar is fitted with a Vetag positive train identification unit which sends information such as formation, destination and timekeeping to lineside units for feeding into the station displays. Visually, each half of the Metrocar is more or less a mirror of the other except for the fitting of the Brecknell–Willis high reach pantograph on the roof of one vehicle to collect the 1,500v DC

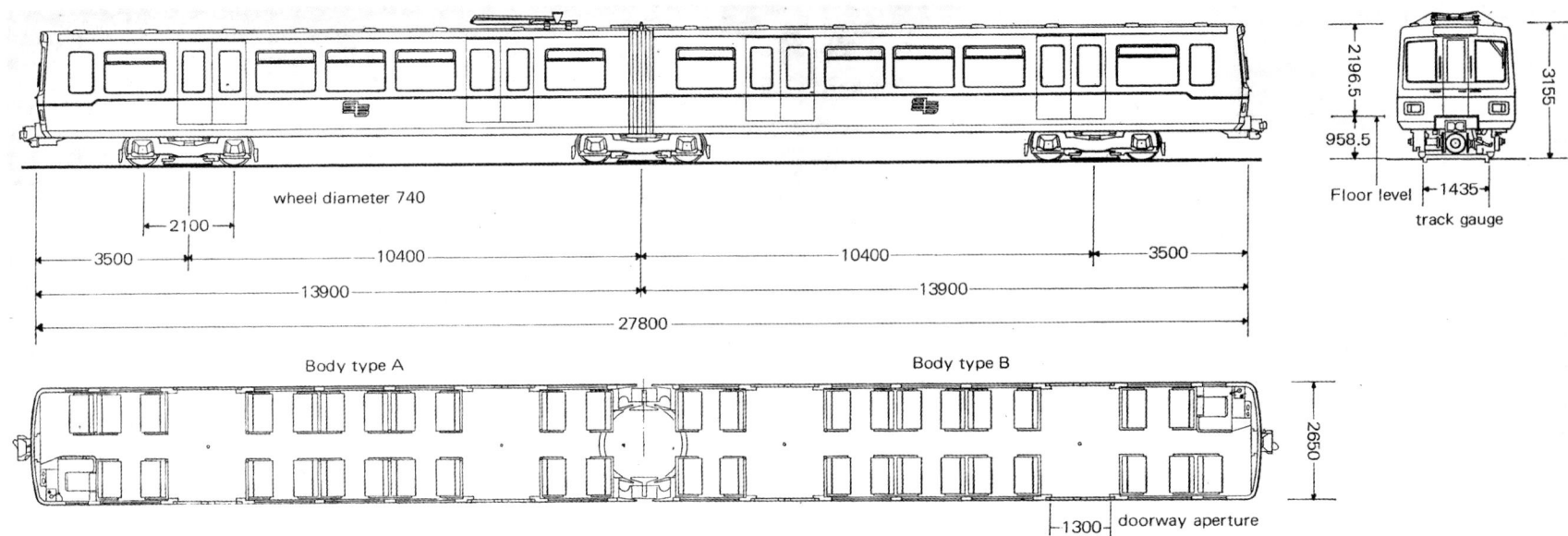

Diagram of the layout of an articulated Metrocar. (Author's Collection)

from the overhead catenary; again this is a tried and trusted unit, having seen several years use on British Rail before the introduction of the Metrocars.

The units were numbered 4001–4090; there are no individual numbers for each half of the set as they are considered permanently coupled as one vehicle. One unusual fact is that they were not built in numerical sequence. As mentioned the two prototypes, 4001 and 4002, were delivered first, and then the first production set 4003 arrived next in August 1978 followed by 4004, 4005 and 4006 in October of the same year. The next sets delivered were 4034, 4009 and 4050, and after that the cars were numbered in no order whatsoever. This was brought about by the fact each Metrocar is numbered based upon the serial number of the electrical equipment case that was made by GEC and delivered to Washwood Heath ahead of construction. The cases were put into store and taken out in a random order when needed, hence the running numbers do not equate to build order. The last built set was 4030, being outshopped from Washwood Heath in November 1981.

The choice of livery for the Metrocars was based upon that given to the PTE's buses. It comprised a shade of yellow, branded Newcastle Cadmium Yellow on the lower body side, with the upper body finished in brilliant white. Between the two was a French Blue stripe. The roof was painted Storm Grey whilst the front window surrounds were Matt Black; letters, numbers and logos were black vinyl cut stickers. All the underframe equipment and bogies were painted gloss black; the prototype cars had an earlier variation of the livery when delivered with maroon used for both the body side stripe and the equipment cases and bogie frames. The two prototypes were not compatible with the production builds after they arrived at South Gosforth and spent several years in store before being sent to Hunslet in Leeds in July 1984 to be rebuilt to match the rest of the fleet, including changing the livery to that of the production cars. It would not be until July 1987 that these two sets finally entered passenger service.

One of the problems during construction was that cars were being delivered at a faster pace than the South Gosforth depot could house and commission them. Several units were delivered to the Tyneside Central Freight Depot in Gateshead during 1979 and 1980 to be stored pending transfer to South Gosforth. Unit 4077 was delivered to the National Railway Museum at York in May 1979 and displayed there until moved to South Gosforth in November 1979. An arrangement had also been made to store vehicles on a section of the closed Woodhead line near Sheffield but this option

The car sheds at South Gosforth had to be responsible for the maintenance of the Metro fleet and were equipped to carry out all nature of heavy overhauls. Here Metrocar 4036 has been lifted off its bogies by the depot's hydraulic jacks. (Tyne & Wear PTE)

General view of South Gosforth car sheds with Metrocars on depot either awaiting maintenance or their next duty. In the centre of the image is one of the Metro's locomotives used for engineering trains. (Russell Graydon)

was never needed. When the formal opening ceremony was conducted on 6 November 1981 it was unit 4020 that had the honour of conveying HM Queen Elizabeth II from Monument to Heworth with unit 4085 also in the consist. By December 1982 all eighty-eight of the production units had been delivered and commissioned and it was a case of them settling into their new life moving millions of people around the Metro network.

In the first year of full operation the Metro was responsible for 60 million passenger journeys. Ironically this has been its most successful year to date, given there would have been a number of special journeys made in those first few months by locals and some from further afield wanting to sample this new fangled railway, but the numbers have remained good in the following years and the network has lived up to both expectations and its huge price tag of £288 million. No one today doubts that it was worth the expense.

The idea of integrated public transport, of which the Metro formed a huge part along with local buses and main line trains, was set back by the 1985 Transport Act which led to the deregulation of local bus services and placed them in the hands of private operators instead of the PTE. These firms could run whatever services they wanted and charge their own fares; many set up in direct competition to the Metro which went against the original ideals for complementary bus routes. The Metro had been designed with buses feeding into it and in return it fed passengers from the trains to the bus, with the two not in direct competition with each other. That idea, however, was perhaps far too visionary and utopian for the 1980s when public ownership was shunned in favour of private enterprise and market forces.

Three new stations were added to the network before the opening of the extension to Newcastle Airport. Kingston Park (between Fawdon and Bank Foot) opened on 15 September 1986, with Pelaw on the South Shields line opening the following day. On 19 March 1986 Palmersville (between Benton and Shiremoor) was added to the network and Newcastle Airport was reached on 17 November 1991.

The Metro settled into the 90s but it was not too long before a re-livery of the trains occurred. During the mid 90s the entire fleet was painted in either a red, blue or green colour scheme with the cab ends in yellow. The black cab front window surrounds were retained and the yellow extended back from the cab to nearly the first set of doors but was angled at 45 degrees so the yellow tapered out as the angle swept back. This livery had to be modified a decade later to comply with the Disability Discrimination act that required the passenger doors to be easily discernible from the body of the train. This resulted in the doors being painted the same yellow as the cab fronts.

Tyneside Metrocar No. 4025 is seen in original livery at South Shields on 24 January 1987. It is heading for Bank Foot, which was the north-west extremity of the network at the time. (Graeme Phillips Collection)

Tyne & Wear Metro cars Nos. 4063 and 4069 (rear), both in the red version of the revised livery, leave Heworth, near Gateshead, with a South Hylton to Airport service on 16 January 2013. (Allan McKever)

Metrocar 4044 and partner pause at Whitley Bay with a coastal service on 6 September 2019. This grey/black/yellow colour scheme is expected to be the final livery the cars will carry until they are withdrawn. (Graeme Gleaves)

Metrocar 4043 at the front of a twin formation calls at Tynemouth with a St James to South Shields service on 6 September 2019. (Graeme Gleaves)

A major development that took shape in the late 1990s was the extension of the system to Sunderland. This echoed the plans of the NER who had investigated this option shortly after the North Tyneside lines were electrified. The extension was to run from the junction at Pelaw for 18.5 kilometres to South Hylton via Sunderland. The cost of this would eventually come in at £100 million, with funding coming from Nexus (the brand name for the PTE), the European Regional Development Fund, Railtrack (soon to become Network Rail) and the remainder from central government. The route was to use a mix of route sharing with Network Rail and take over redundant freight lines to give the Metro its own permanent way. The Durham coast line between Newcastle and Sunderland would continue to be operated by Northern Rail trains with the Metro sharing the infrastructure in a unique arrangement. This meant that all ninety Metro cars had to be registered to run over Network Rail metals and entered onto the TOPS (Total Operating Process System) rolling stock system. They were given the TOPS class of 994 and entered onto the rolling stock library as units 994 001–994 090, the last three digits of their Metro running numbers corresponding to the last three digits of their TOPS number. Between Sunderland and South Hylton the route is exclusively used by the Metro.

The extension opened on 31 March 2002. Intermediate stations between Pelaw and Sunderland were provided at Felgate, Brockley Whins, East Boldon, Seaburn, Stadium of Light (for the AFC Sunderland football stadium) and St Peter's. The first three were already part of the existing main line and are still shared with Northern Rail services, while the latter three were new constructions. Stations were provided beyond Sunderland and named University, Millfield and Pallion. The official opening ceremony for the new extension was conducted by HM Queen Elizabeth II on 7 May 2002. Prior to this event another station had been added at Park Lane on 28 April 2002, located between Sunderland and University. Elsewhere on the network two new stations were added during the first decade of the twenty-first century. Northumberland Park is on the north side of the former NER Tyneside loop between Shiremoor and Palmersville, and was opened on 11 December 2005. The latest and to-date final addition to the network was Simonside station, situated on the South Shields line between Bede and Tyne Dock; it opened for business on 17 March 2008. Surprisingly the extension did not require any additional stock to be built. The service on the expanded network was operated using the same ninety Metrocars they started with. Some services were reduced to two-car formation whilst the testing and training was being undertaken prior to the extension opening, and by improving maintenance and stock diagrams the system can operate with the same fleet.

The management of the network has changed during the twenty-first century. The Tyne and Wear PTE set up the Nexus brand as a trading name for its operations that included both the Metro and the Shields Ferries. In 2010 Nexus put the operation of the Metro out to tender. The winning bidder was DB Regio Tyne & Wear Limited, a company set up by Arriva UK Trains. They ran they network as a franchise until 1 April 2017. Upon the expiry of this contract Nexus did not re-let the franchise or invite any fresh bidders as they had decided to once again take over the running of the system.

The fleet of Metro cars underwent a refurbishment programme at Wabtec in Doncaster between June 2011 and July 2015, with eighty-six of the fleet being so treated. The work involved corrosion repairs to the bodywork, new seating and new floors, along with improved lighting. The space available to wheelchair users was enlarged along with 'call for aid' alarms being fitted at these points to alert the driver of an emergency. The units that were not refurbished were prototypes 4001 and 4002 along with 4040 and 4083. Units that had been refurbished were painted in a new colour scheme of dark grey body with black window surrounds and yellow passenger doors; the cab front is similarly dark grey with cab front windows having a black surround. A yellow stripe covers the full height of the corners of each cab end and small square yellow panel extends from below the central front window down to the coupler and is a third of the width of the cab front. The Metro M logo is in black on this panel and again against a small yellow panel midway down the body of each car. Car numbers are

Above left: **Car 4073** calls at South Gosforth station with a South Shields to St James service via the coast. The large building on the platform is the Metro control centre. (Graeme Gleaves)

Above right: **Close up** of the coupling arrangement between two Metrocars. The pipe on the top of the coupling carries the main air supply between the sets. Below that is the mechanical coupling that keeps the cars joined together and below that are the electrical connection boxes that transfer control signals between the vehicles, ensuring they operate in synch under the control of a single driver. (Graeme Gleaves)

in white carried below the head and marker light cluster on the driver's side of the cab front and again above the first large body side passenger window below the orange hazard cantrail stripe. This uniform livery put an end to the numerous special liveries members of the fleet carried. Unit 4001 had been repainted into its original livery but lost it when it was repainted into the new colour scheme in 2017. The list of special liveries carried by the Metro cars is extensive and too long to cover in full here. They have been applied to either advertise companies or mark special events, such as 4032 which carried a special gold livery to mark the Queen's golden jubilee in 2002. The process had begun in 1989 with Metrocar 4044 painted yellow and named 'The Director' to mark the 150th anniversary of the Brandling Junction Railway. It was joined the same year by 4051 which was painted claret and name 'Times' to mark the 150th anniversary of the Newcastle and North Shields Railway. Metrocar 4080 was painted white in 1990 with butterfly motifs to mark the opening of the Gateshead Garden Festival, although it was car 4087 that was transported to the event and kitted out as a Metro

Close up of the pantograph of a Metrocar. (Graeme Gleaves)

Travelcentre whilst on display there. The first commercial livery was in 1991 when Metrocars 4054 and 4058 were adorned with cartoon images for the Metroland Indoor Theme Park at Gateshead. Since then many more liveries have appeared. The last one running was 4083 which carried a livery promoting the Emirates airline.

With annual passenger numbers now exceeding the 40 million mark the system is considered in good health but the replacement of the Metrocar fleet is on the horizon. Nexus has invited tenders for this process, and estimates put the cost of a new fleet in the region of £550 million. Several ideas have been put forward for the concept of the new fleet, including the reduction from four to two-car working. The option of having dual supply of not only the Metro standard of 1,500v DC but also 25kV AC as employed on the national network could give them greater flexibility to operate over more lines. No decisions have yet been made and the existing fleet will have to continue in front line service until 2023 at the earliest.

The Tyne and Wear Metro paved the way for the re-invention of the suburban electric railway. Other networks have followed its example in taking over former 'heavy rail' lines and are

Car 4026 calls at West Monkseaton with a St James to South Shields service. The difference in roof details between the two halves of each Metrocar is evident due to the need to mount the pantograph. (Graeme Gleaves)

ating the 100th anniversary of the introduction of Britain's first all-electric suburban rail service by the North Eastern Railway, 1904

The Metro issued this commemorative postage pack to mark the centenary of electric trains on Tyneside in March 2004. The envelope features images of NER, LNER and Metro trains and was carried on the Metro before being posted. (Author's Collection)

running fast, frequent services over modified infrastructure deep into the heart of the city. Manchester's Metrolink and the Croydon tramlink were the first of this new generation that owe their existence to the pioneering work of the Metro, although they have taken the concept further by incorporating tram style on-street running instead of tunnels under the city, an idea that has even been suggested for a future phase of the Metro. The future shape of the network may be open for debate but what is certain is that the suburban electric train has made a triumphant return to the railways in the North East of England, where the passengers never look at a timetable as 'there will be one along soon'. Charles Mertz would be very happy that his vision lives on.

Chapter 14
OTHER NORTH EASTERN RAILWAY ELECTRIFICATION PROJECTS

It can be argued that few other pre-grouping railway companies embraced railway electrification as enthusiastically as the North Eastern Railway. This was due in no small part to the influence of the company's final Chief Mechanical Engineer (CME), Sir Vincent Raven. He was appointed to the post in 1910 having previously served as Assistant Mechanical Engineer to the then CME Wilson Worsdell from 1902. Raven was a well-educated man and had worked for the NER since leaving school, joining them as an engineering apprentice. His enthusiasm for electrification was not just limited to the trains themselves. He saw the benefits of introducing machinery powered by electricity and presented a paper on electric cranes during one of his many visits to the United States. It was in America that he saw first hand the spread and success of the new electric railways working there. Raven would have no doubt played a key role in the introduction of the Tyneside electrification of 1904 and was certainly a key player in the small scheme that was abridged to it: the electrification of the Quayside branch. This was a short branch that left the main line at Manors station, where Trafalgar Yard was located for the marshalling of traffic to and from the branch. It then headed south, down a 1 in 27 gradient and through a section of mostly tunnels to emerge at Quayside Yard. The route was semi-circular in shape and used for freight traffic only. The line was worked intensely due to the gradient dictating limits to the lengths of trains that could safely descend the line. Multiple trips had to be made with shorter trains to service the traffic needs.

There can be no better example of electrification improving the quality of the working environment than the electrification of the Quayside branch. When the line was steam-hauled the engines would be working hard to lift their train up the 1 in 27 gradient, which is amongst the steepest gradients that could be found on a locomotive-worked line in Britain. A steam locomotive that is working hard to haul a loaded train up an incline is going to need to raise a lot of steam in its boiler and that will require the fireman to shovel hard to keep feeding the hungry fire. All the time the loco is drawing air across the fire and the heat inside the firebox intensifies, as does the production of smoke and soot, not to mention the temperature, not just from the burning coal but from the steam it creates. The loco crew cannot escape this environment and to make it even worse the whole process is going on inside the confines of a tunnel that extends the whole length of the line. The return trip is more a case of using the locomotive to hold the train back as it tries to accelerate down the incline, but you are running through the same tunnel you filled with soot and smoke a few minutes ago with the last assault on the gradient. Was there ever a line that needed electrification more than this? The NER made the specification for the locomotives to work this line in a tender dated July 1902. This called for a loco that could work six trains every hour on the line and also be used for shunting in the yards at each end. The loco would need to be able to haul a trailing load of 150 tons up the line at a speed of 10 mph. Two locomotives and a quantity of spare parts were required. The specification also called for multiple unit operation between the two locos and the provision of air operated sanding and brakes. The brakes had to act not only on the wheel surfaces but also the flanges to ensure it could stop on the steep gradients with ease. Both locomotives would need to be delivered to Newcastle by 31 December 1903.

The preferred bid for the construction of the locos was won by the British Thompson-Houston Company Ltd. They sub-contracted the supply of the frames, bodies and mechanical parts to Brush Electrical Engineering and it was at that company's works that the two locos were assembled. What emerged in early 1905 was a Bo-Bo locomotive that, like the NER's electric passenger stock, owed a lot to the American influence. The design became known as a 'Steeple Cab' due to the driving cabin being mounted centrally on the underframe with an angled bonnet sloping down from just below the cab front widows to curve off the end in a taper at the buffer beam ends. It is inside these bonnets that the air reservoir tanks for the braking and sanding systems were mounted at the buffer beam end; also inside were the banks of traction resistance that were used to accelerate the motors. At each bonnet end, just before it curved down to meet the buffer beam was mounted an electric headlight and either side of this were the lids to fill the cylindrical sandboxes that ran down inside the bonnet. Sand was important on such a steeply graded line and the locos had the capacity to carry plenty of it. Standard NER pattern loco buffers were fitted, with drawhooks and three link goods couplings. When built the locos had train air brake swan neck connections either side of the draw hook but the requirement for multiple unit operation as set out in the specification was dropped at some stage as the locos never had this equipment fitted. The central cab had a sliding door on each side for access and controls were provided to drive the loco in each direction. Three lamp irons were fitted above each buffer beam and a fourth one was fitted to the cab front between the two cab windows. A single air whistle was mounted on the centre of the roof.

The locos were mounted on bogies with 3 foot diameter wheels on a 6 foot 6 inch wheelbase. Each axle was fitted with a 160hp traction motor. Between the bogies and mounted in a frame underslung from the loco's underframe was an electrically driven compressor which charged both air reservoirs in the bonnets to supply the brakes, sanding gear and whistle. The collector shoes were, like on the passenger stock, mounted on a wooded beam that ran between the axle boxes on each side of the bogie, and this was fitted to both sides of both bogies.

Third rail was used in the tunnel section of the line but in the yards at either end concerns for staff safety meant that the electrification was provided by overhead trolley wires. This meant the locos had to be fitted with bow collectors to connect with the wires. These were mounted on a wooden beam transverse of one of the bonnets and were sprung loaded. In operation it was the duty of the fireman to lower the bow collector before entering the tunnel and the driver had to use a pair of knife switches in the cab to change between overhead wire and third rail operation; needless to say there were a few instances of the bow collector being damaged when it was either late being lowered or not lowered at all! The locos were 37 feet and 11 inches in length over buffers and weighed in at 56 tons. One has to remember that when built in 1905 they were amongst the very first main-line electric locomotives in Britain. Early deep level tube railways that ran under London had been employing small electric locos since 1890 but the use of an electric locomotive on a freight train was a first for not only the NER but for the UK. When built the locos proudly wore the company livery of lined green and were numbered as 1 and 2 respectively.

Electric operation commenced on the Quayside branch on 5 June 1905 and one can only imagine how different the working conditions became for the loco crews from that point onward. The shunting further down the quayside continued to be carried out by a steam loco and it was one of the duties of the electric locos to take the steam engine down the line in the morning and haul it back at the end of the days shift. Trains on the branch were never hauled down the incline, instead the loco would

One of the two NER Bo-Bo locomotives pictured early in its life under the wires in the Quayside yard. The original bow collector has gone and has been replaced with the roof mounted pantograph, which dates the image to sometime after 1918. (Author's Collection)

push the train and hold it back from running away with its loco brakes. At the other end of the train was a specially modified brake van which had sandboxes of its own. This van also used its handbrake to check the descent of the train and was uncoupled at the bottom of the line then run off into a spare siding by gravity so the wagons could shunted. The van would be picked up by the formed-up train ready for the return trip up the branch. The locomotives were more than up to the job and could handle trains of up to 140 tons in weight. Both were maintained alongside the electric passenger stock at Heaton Car Sheds. When that depot was destroyed by fire the responsibility

Pictured some time close to the grouping of 1923, this image captures one of the locos about to haul a load up the incline from Quayside to Trafalgar Yard. The loco wears a much simpler livery than the previous image. (LCGB/Nunn Archive/Author's Collection)

LNER ES1 class Bo-Bo electric locomotive 26501 at Heaton (52B) shed on 26 June, 1960. (Ron Fisher)

shifted to South Gosforth depot and this is where they remained up to the end of their working life on 29 February 1964, when the operation of the branch was handed over to BR/Drewery Class 03 diesel-mechanical locomotives.

The locos underwent several minor modifications during their working life. The crews, who had been roasted, choked and poached when the line was steam-operated complained the electric locos were too cold on early shifts, so they were both fitted with cab heaters at the end of 1906. The bow collectors were replaced by pantographs mounted on the cab roofs sometime after the First World War. Because of this the whistle had to be moved to a new position on the cab front adjacent to the lamp bracket but as the locos operated in each direction a second whistle had to be added later on to the front of the other end of the cab. Locomotive No. 1 was modified just before nationalisation whilst No. 2 was not modified until it was in the ownership of British Railways.

The handrails that were bent to contour around the bow collector beam were not altered when the beams were removed and it was not until 1948 that both locos got straight handrails on the edge of

Loco 26500 photographed from a passing train at Trafalgar Yard, near Manors station, in 1963. (Author's Collection)

both bonnets. The buffers were changed a few times, once by the NER and then by the LNER. Both locos were painted black by the LNER with red lining and the letters LN one side of the cab door and ER on the other; during the war this was shorted to just NE with one character either side of the cab door. The LNER renumbered the locos 6480 and 6481 in 1946. Shortly before nationalisation the air brake hoses on the buffer beam ends were removed; this was not surprising as there is no evidence that they were ever used. Only one loco was required each day to work the branch and the other sat at the home depot either undergoing maintenance or available for other work. Given that these were designed as shunting locos with a low gear to provide a higher tractive effort, any additional work they could undertake was limited. Certainly, hauling passenger trains was not in their remit and it was the passenger stock that had air brakes on the NER. The locos could have been used to shunt stock at the car sheds where the central cab would have made them well suited to this work, but they lacked the Cowhead couplers that were a feature of the passenger stock so their use on this duty was limited too.

When British Railways took over in 1948 the two locos were renumbered again to 26500 and

26501. Both locos had the British Railways lion and wheel emblem applied to their black paintwork in 1952 and remained like this until June 1961, when 26500 was repainted in fully lined NER green; 26501 was similarly treated in March 1962. The numbers were carried to the left of the cab door and the second style of lion and wheel emblem was mounted above it; to the right of the cab door the NER coat of arms was applied. This was a fitting way for them to see out the last two years in service.

After withdrawal both locos were put in store at South Gosforth car sheds but moved in January 1965 to Hellifield shed. It was from here that 26501 was sold for scrap in April 1966 and returned to the North East to be broken up at Choppington. Loco 26500 went on to be stored at AEI in Rugby before going on display at Leicester Stoneygate museum. When that museum closed in 1975 the loco was moved to the National Railway Museum in York. It was restored to NER livery and placed on display. The National Collection opened a new facility called Locomotion at Shildon in October 2004. NER no.1 (26500) was chosen as one of the exhibits to be moved there and remains there to this day, very appropriately displayed in its native North East.

Vincent Raven succeeded Wilson Wordsell as Chief Mechanical Engineer of the North Eastern Railway in 1910. He now had free reign to indulge in projects in his own name and did so. Despite being innovative in the application of early electric traction one must remember that the NER was mostly powered by steam and Raven was responsible for adding superheaters to many of the NER locomotives to make them better performers. The Chairman of the NER, Alexander Butterworth, asked him to look into further electrification on the NER with the consultants of Mertz & McLellan. Both Vincent Raven and Charles Mertz visited the United States in accordance with this request to see the extent and nature of electric railways operating there. In America they would have seen not only the urban electric networks but also the main line routes that had started to electrify. The US had a large head start on the UK in terms of railway electrification and had quickly become the world leader on this form of traction. Raven returned to England and presented a plan to the board in January 1912 that advocated the NER adopting

Both locos photographed in late 1964, after withdrawal, at South Gosforth depot. (Author's Collection)

The beautifully restored NER No. 1 (BR 26500) on display at the National Railway Museum site at Shildon on 27 June 2019.
(Graeme Gleaves)

Close up view of the bogie of NER No. 1 showing the collector shoes hanging from the shoe beam which is supported between the axle boxes. The box above the beam contains the shoe fuse that was designed to protect the loco against a surge in traction current.
(Graeme Gleaves)

The underslung compressor on NER No. 1. This machine was nothing more than a motorised bicycle pump and was operated by air pressure and electricity to keep the locomotive's air reservoirs topped up to enable braking and sanding. (Graeme Gleaves)

electrification of its trunk routes and the use of electric locomotives, not steam, to haul the trains. The figure of £1.5 million was put forward as the estimated cost of the conversion to electric traction. The NER board considered the proposals and concerns were voiced about the accuracy of the estimated cost, as the board considered the final bill would be much higher. Butterworth was not convinced that the scheme was viable but did give Raven permission to electrify an eighteen-mile long freight line to act as a small-scale test bed for his proposal. The line chosen ran from Shildon, a few miles south of Bishop Auckland and headed more or less due east to the Newport Yard on Teesside. No passenger services ran the entire length of the line but parts of it were used by passenger services whose lines intersected the route. The sole purpose of the line was the transport of coal and coke from the yard at Shildon to Newport, where it was distributed to the blast furnaces and ironworks in the locality, and also onward to the area's docks for shipment to other parts of the UK and overseas. A section of the route dated back to the original Stockton and Darlington Railway opened in 1825. There was a steady flow of traffic and parts of the route were quadrupled to accommodate the many trains using it. It was also convenient that the line was mostly downhill heading towards Newport, as this was the direction

Official portrait of Sir Vincent Raven. (The Engineer)

The electrification work was authorised in 1913 and the system of electrification chosen was an overhead wire with the wires energised at 1,500v DC. The negative return of the traction current was to be by the running rails. Overhead wires were made from copper and supported by steel catenary masts of various types depending on the location. The wires were set 16ft 6 in above the running rails but this was increased by 2 feet where the line passed over a level crossing to provide plenty of clearance to large loads using the roadway to cross the railway line.

Electric power was supplied to the system directly from three power companies: The Newcastle-upon-Tyne Electrical Supply Company Ltd, Cleveland and Durham Electric Power Company and Cleveland and Durham Electric Power Ltd. Two substations were built to feed the line: one at Aycliffe which obtained a 20,000v AC supply and used a rotary convertor to step it down and convert it to the traction voltage, whilst the other substation was built at the very end of the line at Erimus near Newport, where a 11,000v AC incoming supply was dealt with by another rotary convertor. An electric control room was provided at Newport and section switching equipment was

the heavier loaded trains travelled in, meaning the lighter empty trains would face the uphill return leg.

NER Bo-Bo locomotive No. 6 at work on a heavy load somewhere between Shildon and Newport during the early years of the electric service. (Author's Collection)

installed at signal boxes along the route and could be operated by the signalmen as directed by the Newport control room.

The order for ten locomotives to work the line was placed in May 1913 and Darlington Works was given the task of building them. The locos were designed by Vincent Raven and are best described as a much enlarged and 'boxier' version of the two Quayside locos. Like those they were Bo-Bo design with two bogies, each with two powered axles. The wheels were 4 foot in diameter and each axle had a 275hp traction motor supplied and fitted by Siemens Brothers, who supplied all the electrical equipment for the locos. The locos featured a large central cab with a bonnet either side but these bonnets had a much gentler slope than the quayside locos and ended in a square front above the buffer beams; immediately above the buffers was a slatted grille that served to admit air to cool the resistance banks housed in the bonnet.

The bogies carried sandboxes at their leading end and midway point that delivered sand under only six of the eight wheels regardless of direction of travel to aid traction, and was applied under pressure from compressed air. Leaf springs supported the axle boxes and unlike the Quayside locos the buffer beam was affixed to the front of the bogie and not the loco's body; standard goods buffers were fitted and three-link couplings. The inner ends bogies almost met under the cab so there was no room to mount the air compressor there, instead it was fitted inside one of the bonnets.

Inside the cab the driver's position was on the right-hand side of each end and a power and brake controller handle was provided, along with an ammeter and pressure gauges for the air brakes, these acted on the loco but train brake pipes were provided on the ends of each bonnet so it could operate air-braked stock if needed. The large cab also featured a handbrake wheel in the centre and a pair of traction motor blowers that fed air over the motors to keep them cool when running under power. Three lamp irons were fitted along the front of the bonnet above the buffer beam and a fourth one added at the centre on top of the bonnet. The roofs of the cab had a very slight curve as they were where the two pantographs were fitted. These were raised and lowered by compressed air and a special key had to be inserted into a switch to keep them raised. This key also opened the high voltage control box so it served as a safety feature that the high voltage equipment could not be accessed with the pantographs raised. It was normal practice to operate the locos with both raised at all times but in the event of a failure of one pantograph the locomotive could continue to operate. A hand pump was provided to build up sufficient air had the locomotive's own air reservoir gone 'flat' after a period of being switched off. One final feature worth mentioning is the brass whistle that was fitted on each cab front between the windows; this was powered by compressed air and operated by a plunger at the respective end's cab control position.

Nine of the locomotives were ready to be placed in service by December 1914 and emerged from Darlington painted in plain black with red lining. The legend NORTH EASTERN was split either side of the cab door, and on the front of the bonnet each locomotive had an oval plate bearing its running number. The numbering of these ten locos followed on from the Quayside ones as the NER had a separate sequence for electric locomotives up until 1914, but from the delivery of these locos there was just one loco numbering sequence. The Shildon locos became Nos. 3–12 and any steam locos that clashed with those numbers underwent renumbering to make way for them. The first eight complete locos were sent north to Shildon in June 1915 and test running began that month, being concluded quickly to enable a ceremonial first train to be hauled over part of the route on 1 July 1915 by locomotive No. 3. The full length of the line was worked by electric traction from 10 January 1916. The former steam roundhouse number 3 at Shildon had been converted to provide the locos with their own dedicated shed where all servicing and maintenance was carried out. Heavy repairs would be carried out at their birthplace of Darlington Works. Loco No. 11 was completed and sent to Shildon during the second half of 1915 but the last loco to be completed, No. 12, was not fitted out until December 1919. The electrical equipment for this loco, being supplied by a German firm, was delayed due to the outbreak of war.

In use the locos proved to be both reliable and more than up to the job. The maximum load allowed for a train hauled by a single loco was 1,000 tons but this was raised to 1,400 tons in November 1922.

The locos came under the LNER from 1 January 1923 but this was one line where the handover was hardly noticed. Locomotive No. 6 was repainted by the LNER very early into their ownership following collision damage sustained during the last days of 1922, when it failed to stop whilst hauling a heavy train and had collided with a train of empty wagons at a speed of approximately 3mph. The repaint was again in black with the legend L&NER on the cab side and a slightly smaller brass number plate fitted to each bonnet end. All the locos got similar treatment over the next ten years but only No. 6 bore the '&' character, the remainder had just LNER applied. Loco No. 5 was unique in that it never got the LNER lettering applied and also kept its NER numberplate. It remained in this condition until scrapped in 1951.

The route saw a decline in traffic from the start of the 1920s and there was barely enough work to keep half the fleet busy by 1925. It was during this year that loco No. 9 took part in the Stockton & Darlington Railway centenary procession and was displayed at Faverdale for ten days with several other exhibits from the event. In 1928 Nigel Gresley put forward a proposal to convert one of the locomotives to an experimental diesel electric by fitting a Beardmore 1,000hp engine driving an English Electric generator set. The conversion would have involved the rebuild of the body work to a design with a cab at each end of the structure and the engine and generator mounted centrally. The intention was to put the loco to work on coal traffic in the Peterborough area but the conversion was cancelled after the engine manufacturer expressed doubts their unit would be up to the task. By the mid 1930s the electrification equipment on the line was due for an upgrade, but the LNER decided not to proceed with this given the very limited use it would have for the cost needed. The depression had hit traffic even further and by this time there was probably only enough work for two of the ten locos each day. The remaining traffic over the route went back to steam locos in January 1935 and Shildon shed was closed. The locos were moved from Shildon to Darlington for storage in July 1935. They were only twenty years old and had a very good record for reliability, although this could have been due in part to the amount of running they had put in over the previous two decades being far lower than other locos of an equivalent age. No doubt the LNER considered there may be further use for them yet, though at that time the only other line that used the 1,500v DC overhead wire system was the Manchester South Junction and Altrincham line that the LNER jointly operated with the LMS. This was only eight and a half miles long and was for suburban traffic; there was no scope for a locomotive designed to haul heavy freight trains at 25mph.

For nine of them Darlington was to be their home for the next twelve years; the loco that escaped and found a second lease of life was No. 11. In 1936 it was selected to go to Doncaster Works and undergo a conversion to enable it to work on the Woodhead line that was to be electrified at 1,500v DC. The LNER wanted new-build locos for the passenger and freight traffic but had identified a need for banking locos to assist heavier trains up the inclines that were a feature of the route. English Electric were contracted to provide the electrical upgrade for NER No. 11, but the commencement of the work was delayed by the start of the war and it was not until 1941 that No. 11 reached Doncaster for the work to begin.

The rebuilding included new motors and a new pantograph. The new motors raised its horse power from 1,100 to 1,256; the new pantograph was mounted in the centre of the cab roof and there was only one fitted. Extra sandboxes were added at the inner ends of both bogies so the loco could sand all wheels in both directions. The cab doors were moved to the right-hand side of each cab adjacent to the driving position. I have heard that the driving position controls were moved from the right to the left-hand side of the cab but have not seen evidence to support this, although it would make sense to bring the cab layout in keeping with other locomotives at the time. The alterations to the bonnet included the removal of all four lamp irons and their replacement with electric marker lights in their place. A single lamp iron was fitted on each end below the upper headlight and the handrails on each bonnet were modified to accommodate this change. The grilles on the bonnet fronts to ventilate the interior were removed and replaced with long lozenge-shaped air intakes on the upper side edges of each bonnet. The loco was finished in all-over black and carried the wartime company logo of just 'NE' mid

NER electric loco No. 4 stored in Shildon shed in company of some assorted locomotives and stock. (Rail Archive Stephenson)

Former Shildon–Newport loco No. 11 pictured carriage shunting at Ilford depot in its final guise as BR 26510. (LCGB/Nunn Archive/Author's Collection)

height and midway along the cab sides, with the number 11 below it; the brass number plates were removed as part of the nose end alterations. The rebuilding was completed by December 1942; there was no urgency on the project as electrification work on the Woodhead route had been completely halted due to the war effort. The intention was to convert the remaining nine locos to the same specification and when the LNER introduced a new classification system in October 1945 for locos the ex-Shildon locos were given the classification of EB1 (Electric Banking type one); at the same time the two Quayside Branch locos became ES1 (Electric Shunting type 1). The following year all ten

were renumbered, in build order, as LNER 6490 to 6499. All but No. 11 were in store in the Darlington Works paint shop at this time so were not really affected by the new classification, but No.11 was an ongoing project, despite being stored at Doncaster awaiting resumption on the Woodhead line works. It received its new running number of 6498 on each cab side below the LNER lettering. All ten locos were transferred to storage at South Gosforth car shed during 1947 and remained there during the transition from the LNER to British Railways.

After nationalisation, 1949 was to be a year that saw a downturn in the locos' possible future. The Woodhead line banking project was cancelled so all nine of the unconverted locos were reclassified as EF1 (Electric Freight type 1), though the prospects of them ever hauling a freight train again were by now quite remote. They were also renumbered by British Railways, becoming 26502–26511, again in build order. These numbers were applied whilst in store either on the side of the right-hand bonnet or on the central side of the cabs. The solitary EB1, 26510, was released from store, bearing its new number below the first type of lion and wheel British Railways logo. It was towed south to Ilford depot where a new role had been found for it as depot shunter for the newly electrified suburban lines out of Liverpool Street. The 1,500v DC electrification went as far as Shenfield and Ilford was the depot that housed all of the multiple unit stock that worked the route. Meanwhile at South Gosforth a last ditch attempt was made to find some further use for the EF1s. One loco was towed to the Quayside branch to see if it would fit the line's clearance. This was reported as happening on 11 May 1950. The Shildon locos would have needed extensive conversion to make them compatible with the lower voltage operation of that line, although their larger size would have enabled them to work longer trains on the branch. In the end the experiment was clearly not worth pursuing and all nine were officially withdrawn from stock as of 21 August 1950. Eight of the locos were sold to Messrs Wanty of Catcliffe and broken up by them in their yard near Rotherham. The ninth loco, 26504, was taken to Darlington works where its bogies were recovered as spares for the loco sent to Ilford. The rest of 26504 was scrapped on site at Darlington.

Loco 26510 remained in use at Ilford and underwent another change of identity in 1959 when it was transferred from capital stock to departmental stock; the change had no effect to the nature of the work or location it was based at but it did result in the loco bearing the legend DEPARTMENTAL LOCOMOTIVE 100. It continued in this guise until November 1960 when the suburban lines out of Liverpool Street along with Ilford depot were converted from 1,500v DC to 25,000v AC overhead electrification. The ex-NER locomotive was not converted to work from the new supply and was withdrawn from use, spending four years stored at Goodmayes Yard before being towed to Doncaster works where it was scrapped on site.

The Shildon to Newport electrification had been planned as a test bed but events prevented it from being properly evaluated. The First World War was underway when the locos hauled their first trains and by 1922 the NER was heading towards the grouping that would create the larger LNER.

Vincent Raven had experimented with another electric locomotive in 1920 when the tender of a J class 4-2-2 steam loco that had been withdrawn in 1919 underwent rebuilding into an experimental six-wheeled electric locomotive. It is believed no photos of it exist, but drawings and first hand testimony to its existence do and they show that the locomotive was fitted with a single motor inside the body that was vertically mounted and drove the centre axle by bevel gearing. The loco was reportedly trialled using third rail collector shoes between Jesmond and Gosforth and there were reports of it being used as a generator when towed on the Shildon line by a steam loco. Whatever its purpose this mysterious beast was scrapped sometime around 1923.

Far less mysterious was Raven's final electric locomotive. He had not given up on the idea of electrically hauled express trains on the East Coast Main Line, or the NER operated section of it at the least. The NER had not made any decision on the proposals that had seen the Shildon to Newport line electrified as a test bed and Raven wanted to build the loco that would be at the front line of the new electric passenger fleet as a show-piece. In 1919 Mertz & McLellan submitted a second report on their proposals for the extension of NER electrification which covered the east coast route via Darlington and the diversionary section via Stockton. The following year Sir Vincent Raven (he had been knighted for his wartime work as Superintendent of the Woolwich Arsenal) published a more detailed report looking at the NER's section of the east coast route from York to Newcastle and the savings that could be made by switching to electric traction using the 1,500v DC system that had been employed on the Shildon to Newport line. He recommended to the board that a prototype electric locomotive be built and was given permission and a budget of £20,000 to do just that. In the summer of 1920 he visited America again to view electric locomotive types in use there and submitted his report in October upon his return.

Design work on the locomotive was completed at the end of 1920 and the order placed with Darlington Works in January 1921. Electrical equipment was to be provided by Metropolitan-Vickers. The locomotive would be unlike anything Darlington Works had produced so far. It shared certain characteristics with the Newport to Shildon locos such as the large central cab flanked by slightly tapered bonnets, but the chassis was a 4-6-4 type (also referred to as 4-Co-4) with bogies at the outer ends, each with a pair of unpowered axles of 3 foot 7 inch diameter wheels. The centre of the chassis under the cab had three axles with 6 foot 8 inch diameter wheels, each with its own 300hp traction motor, giving six in total. The driving wheels were spoked with only six spokes but each of these split into three near the rim of the wheel to increase the spread of track force better around the outside of the rim. The transmission of power from the motor to the wheel was a complex affair using quills, pinions and springs which all interlocked but had room for travel to allow for the movement of the axle's suspension. This meant the motor was held rigid in the underframe and also meant the motor's weight was not directly transferred to the track. Fifty years later a slightly different method of divorcing a motor's weight from the 'unsprung mass' at rail would be a feature of the APT. The reason for this is that at high speed the weight of the motor is exerting considerable track force which can over time damage the permanent way. By placing the motor before the suspension the forces it exerts are greatly reduced. The loco was designed with a top speed of 90mph and was fitted with both air and vacuum train brakes so it could haul any passenger stock operating the east coast route. The loco's own brake was air, and a compressor and air reservoirs were mounted inside one of the bonnets, along with the vacuum exhauster and resistances for traction control.

In the other bonnet was an electric train-heating boiler and the two ends of the loco could be identified easily, as the one with the boiler had no ventilation louvres at the end but did have the boiler filler cap on top. Each bonnet had four hinged access hatches on the side and a pair of long torpedo ventilation intakes on the upper edges. A row of three electric headlamps were mounted above the buffer beam with a fourth in the centre of the upper edge of the bonnet face. The centrally mounted cab had an access door on the left-hand side as you looked at the broadside view with a row of three evenly spaced droplight windows along the rest of its length. The driving position was on the right of each cab. Four huge sandboxes were mounted between the driving wheels, which could sand in each direction either side of the central wheel, and half size ones on the outer ends of the driving wheels between them and the bogie truck. Sanding was operated by air as was the pair of whistles on the front of each end of the cab between the two forward facing windows. On the roof of each end of the cab was a pantograph that was raised and lowered by compressed air.

The loco weighed in at 110 tons and from buffer head to buffer head measured 53 feet and 6 inches. It emerged from Darlington North Road Works in May 1922 and at a cost of just over £27,000, over budget and nearly three times the price of a steam powered 4-6-2 locomotive that it was designed to replace. The loco was painted in works grey for the official photos, with lining and the legend NORTH EASTERN on the sides of each cab. An oval brass number plate was fitted to the upper side of the cab bearing the running number 13, while below that was a larger oval moulding bearing the NER's coat of arms.

In the summer of 1922, still in workshop grey, No. 13 went to the Shildon–Newport line for trial running. The NER dynamometer car along with an assortment of 16 bogie carriages comprised its trailing load. The trials were completed and No. 13 headed back south to Darlington Works and came out again in December to be exhibited in the old station yard at York to the North Eastern directors. I wonder if anyone realised at the time that this loco would never run under its own power again? On the first of January

1923 the North Eastern Railway was no more and the London & North Eastern Railway took over all of its responsibilities. Sir Vincent Raven had not asked to be considered for the post of Chief Mechanical Engineer of the new concern, instead looking at retirement with some consultancy work to keep him busy. The new CME of the LNER was Nigel Gresley; he was not so enthusiastic about electrification so there was no prospect of No. 13 or others like her thundering north at the head of an express train. Gresley would go onto design the world record breaking A4 Pacific for that purpose as steam was still king on his LNER.

No. 13 made an appearance in lined LNER green livery at the Stockton and Darlington centenary

LNER No. 13, looking very smart in its green livery. (LCGB/Nunn Archive/Author's Collection)

NER electric loco No. 13 stored in Darlington paint shop on 25 June 1932. (Rail Archive Stephenson)

cavalcade held on 2 July 1925. It was hauled by an immaculately turned out J71 0-6-0 tank engine. The loco was displayed with other stock at the Faverdale wagon shops for just over a week following the event. For the next few years it remained in store in the paint shop at Darlington Works, emerging every couple of years to be an exhibit at various regional railway events: July 1931 in Middlesbrough, September 1933 in Shildon and closer to home in May 1935 at Darlington Works. Joined by the Shildon–Newport locos from July 1936 the loco shared the same fate as most of them. It too was renumbered by the LNER in May 1946, becoming LNER 6999 and classified as an EE1 (Electric Express type 1). In 1947 it moved along with the ten other stored locos to South Gosforth car sheds and upon nationalisation it became British Railways 26600. The number was applied on the oval moulding that had previously carried the NER coat of arms, and the digits slightly overlapped the space, so the whole number looked awkward on the side of the loco. Quite why the LNER did not dispose of it is not clear, they certainly had no work for it from the minute the LNER came into being, there was never a chance of it performing the type of work it had been designed for and no new work ever looked remotely possible. Perhaps it was out of respect for Sir Vincent Raven that the loco was retained as some sentimental curio. Gresley had got the CME job unopposed because of Raven's decision to step down and Gresley's successor, Thompson, happened to be married to one of Raven's daughters. Whatever the reason, when the LNER gave way to British Railways there was no room for sentimentality and along with the unwanted Shildon–Newport locos No. 13 made one final journey in December 1950 to Messers Wanty of Catcliffe and was broken up in their yard at Rotherham. Its recorded mileage was less than 2,500. How much of that was made powered as opposed to towed remains unknown.

Just as the successful Metro proved that the ideas of Charles Mertz were still valid decades after his death, so too did British Rail do Sir Vincent Raven the posthumous honour of electrifying the East Coast Main Line. Electric trains, powered by 25,000v AC overhead wires ran into Central station and onto Edinburgh from 1991. The stock used was the Inter-City 225 set, with Class 91 locomotive, Mk 4 coaches and driving van trailer operating in push-pull mode at speeds up to 125mph, but at the time of writing the service is being passed over to brand new Hitachi Rail Class 800 'Azuma' high speed sets. Proof that some people were simply ahead of their time.

NER No. 13 was an odd choice for a stamp issued by the Caribbean island of St Lucia but this two-part 60c issue proved they did it!
(Author's Collection)

Vincent Raven's vision of an electrified East Coast Main Line was realised many years later. Here Class 91 Bo-Bo 25kV locomotive 91 115 *Blaydon Races* is pictured at Newcastle Central on 29 January 2020 standing at the rear of a service it will push to Kings Cross.
(Graeme Gleaves)

Appendix
THE PRESERVED UNIT

The purpose of this appendix is twofold: firstly to give you the reader an idea of the varied lives led by the last passenger units built for the Tyneside third rail system after they were displaced; and secondly to act as a personal memoir of the effort that has been involved in saving the last of these units for preservation, a task that has been highly frustrating for long periods of time but is still very much ongoing. The preserved unit is a correct formation British Railways built South Tyneside 2-EPB. It is formed of a Driving Motor Brake (DMB) and a Driving Trailer Composite (DTC) but the two preserved vehicles never ran coupled together on Tyneside. Indeed they never ran together as a complete unit in passenger service in any period of their life; the joining of these two vehicles happened much later. So to recount the story of how they got there we start at the very beginning.

The DMB was from the eleventh of the fifteen units built; it was

Photographed at Shepperton in October 1999, 930 017, formed of a pair of ex-South Tyneside EPB motor coaches, was the last example of north eastern third rail electric stock in operation. The unit was one of many employed in laying sandite on the rails to help combat the wheel slip caused by falling leaves.
(Graeme Gleaves)

Platform 18 at Waterloo and a very clean 5791 is at the head of a four-car formation on a service to Windsor & Eton Riverside. The vehicle nearest the camera is the DMBS of the preserved unit, whilst the location was later rebuilt as Waterloo International Station. (SERA Archive)

Waiting to leave ECS to Waterloo, 5793 stands in Clapham Yard as part of a four-car formation. The vehicle nearest the camera is the DTC of the preserved unit. (SERA Archive)

released from Eastleigh Works on 2 March 1955 and added to NE Region stock on 2 April 1955, carrying the running number of 65321. The DTC was from the thirteenth set and carried the number 77112, being released from Eastleigh Works on 23 March 1955 and added to NE Region stock during the second half of April. Both units performed the usual duties whilst based at South Gosforth and were displaced in January 1963. The unit containing our DMB was accepted on the Southern Region during August 1963 which was after the unit with our DTC, which Southern had accepted the month before. Both units went through Eastleigh Works and were modified to make them more compatible with the Southern

Photographed from a passing train on an unknown date, 930 053 looks relatively clean for a departmental unit as it moves through Wimbledon Park depot. (SERA Archive)

Region's existing 2-EPB fleet. This work involved the removal of both the destination blind box and headcode and taillamp cluster from the front of the cab and the fitting of a two digit headcode box between the cab windows. This was of the type being fitted to latter builds of 2-HAP units along with new builds of CIG units at the time and was therefore smaller than the boxes fitted to earlier EPB stock. Also the cab front had the power train line jumpers removed (these had been removed from the SR EPB units several years earlier). The first-class compartment was declassified and fitted with second class seating devoid of armrests but it did retain the first-class luggage racks with the umbrella rail.

The shoebeams were realigned to match the Southern Region standard and the whistles above the driver's window were removed and replaced with two-tone horns. Below the solebar the motor bogies were converted to the 75mph suburban gear ratio instead of the 90mph express gear ratio they had been built with, and the trailer bogies were upgraded to the Mark 3D pattern, a modification that had been carried out on their Southern classmates a few years earlier. One feature that could not be altered was the double sized guard's brake, and this made the ex-Tyneside units instantly recognisable during their lives on the Southern Region. Surprisingly the seats that were of a different style to those of the Southern Region built units were also retained on all but the last five units to be converted. Those had their seating stripped out and replaced with seats of the same pattern as the native SR units. All the units were repainted as part of their rehabilitation in the standard EMU green of the time. The North Eastern Region had started to apply small yellow warning panels to the lower front of the cabs from late 1961 but these were not applied at Eastleigh and the units were released to traffic in all over green. The units had carried no unit numbers on Tyneside so these were allocated and applied using the Southern Region four digit system at the end of the sequence that contained the 2-EPBs that had been built for that region. The ex-NE units became 5781–5795 with the unit numbers carried at the upper centre of the cab fronts. Our DMB was in unit 5791, which was released to Southern Region service on 8 October 1963, and the DTC (now a DTS) was in unit 5793, which joined the service fleet on an unknown date during September / October 1963. Coach numbers remained the same as they had done since they were built with the exception of the E prefix which was replaced by an S as they were 'Southern' units now. The conversion work took just under two months to complete for each unit and they were then accepted at their new home depot of Wimbledon Park on the South Western Division of the Southern Region, which mostly comprised the ex London and South Western routes. Here they worked alongside the native fleet of 4-SUB, 2-BIL, 2-HAP as well as 4 and 2-EPB units on suburban and semi-fast outer suburban work, although within a few years they would be joined by the new 4-VEP units built for the Bournemouth line electrification and subsequently ordered for outer suburban work. Routes out of Waterloo to destinations including Hampton Court, Guildford, Dorking, Windsor and Reading were covered, including branch line work between Wimbledon and West Croydon that became a preserve of the ex-Tyneside units for several years. Our units visited Eastleigh Works for overhauls periodically: 5791 in September 1967 (when it gained a full yellow end with green livery), December 1970 (when it was painted all over blue), July 1973, February 1976, September 1978 and finally in April 1981 when it emerged in the blue and grey livery it was to carry up to the end of its working life on the main line. Likewise 5793 was overhauled May 1968 (green livery with full yellow end), March 1970 (all over blue livery), September 1972, April 1975, October 1977 and April 1981 when the final livery of blue and grey was applied. Our 5793 was fortunate enough to get a final C6 overhaul in June 1983. The TOPS system was introduced during the late 1960s and covered EMUs too; initially the 2-EPBs were Class 413 but this was changed in 1975 to Class 416 and eventually to 416/2. During their lives, starting bells, loudaphone cab–cab communication equipment and AWS were all fitted. Our unit had coaches that were fitted with different types of AWS, the DTS of 5793 had the older 'Baldwin' system but the DMB was one of the last units fitted out with AWS so got the later 'Simplified' system. Exams and maintenance would be carried out at both Wimbledon Park and Selhurst depots.

Both units had well travelled but unremarkable lives in passenger service on the Southern Region. The whole of the Tyneside EPB fleet came up for scrutiny in 1984 due to both a re-organisation of the traffic demands for the Southern Region in anticipation of new stock being ordered and also as a result of falling passenger numbers. Three units, 5788/5794 and 5795 were taken out of traffic in May 1984 to be dispatched to Fratton depot pending either further use or disposal. Despite this the remainder of the fleet were renumbered in August and September 1984 as part of wider SR renumbering scheme for the region's units. The 57 at the start of each unit number was simply changed to a 62, so 5791 became 6291 and 5793 became 6293. This proved short lived as the remaining twelve ex-Tyneside units were withdrawn from passenger traffic in October 1984, and despite a handful being re-instated briefly for Christmas mail traffic (possibly the larger brake van was useful) that was the end of their front line service. All twelve units made their way to storage at Eastleigh carriage sidings pending a future use for them. It is here that the story starts to get interesting.

All fifteen two-car units started 1985 in store. The Southern Region had a long history of squeezing every last drop of use out of its withdrawn stock and many units had gone on to see an extended life span due to 'Departmental Traffic' needs, a case of units being taken over by a certain department of the 'behind the scenes' railway and used for their purposes, as opposed to the main passenger traffic. This could range from de-icing and sandite units to the movement of stores between depots or even as training vehicles. The ex-Tyneside units were ripe for this kind of re-use for two reasons. Firstly they were in decent condition having all been overhauled within the last four years, and it was only their non-standard layout that had gone against them when the future rolling stock plans were being drawn up. Secondly, the units had been built without blue asbestos, unlike certain other units withdrawn at the same time. This material had been used for decades as a form of electrical insulation against fire but had been found to be highly hazardous to health so it was outlawed in train construction and units found to contain it had to go through an expensive de-contamination process. It is not surprising that of the fifteen ex-Tyneside units only three complete two-cars found no further use and were broken up in the late 80s and very early 1990s. As for 6291 and 6293 they got the call to work for the engineering team concerned with the electrification of the Tonbridge to Hastings line. They both moved from Eastleigh to Strawberry Hill in October 1985 and underwent a bogie swap with 2-HAP units 6007 and 6008. This re-instated their 90mph express gear ratio capability. The units were renumbered into the Southern's three digit sequence for departmental stock: 6291 became 053 and 6293 became 054. The coaches were renumbered into the ADB sequence with our DMB 65321 becoming ADB977505 and our DTS becoming ADB977508. The two units were put to work on test trains, including the first train to travel the length of the Tonbridge to Hastings line under electric power. Included in the test train formation was a test coach called MARS that had been converted from a Class 501 EMU and was used to test both alignment and integrity of the newly installed electrification. Also included was a Class 73 locomotive to provide backup traction power from its diesel engine if there was a problem with the third rail traction supply. The cab of 053's DMB was fitted with push buttons and the associated wiring to enable it to start the engine of the 73 by remote control. The DMB of 054 was similarly treated as it was these two cabs that were always at the front and rear of the test train formation. The electrification work was completed early in 1986 and the two units were returned to Strawberry Hill depot only to be pressed back into action during August the following year on test trains connected with the electrification of the line between Sanderstead and East Grinstead. Unit 053 was modified further in October 1988 to be fitted with a small sandite hopper. This meant the former passenger saloon was stripped of all its seating and a doorway was cut into the bulkhead between the passenger brake and the saloon area whilst the bulkhead to the rear saloon of the vehicle was completely removed and all the seating from the interior stripped, thus making the passenger area now an empty space. The guard's position was retained but the sandite hopper (smaller than those used for dedicated sandite units) was installed in the former saloon along with the associated pumps and control equipment, with delivery pipes passing through the floor to the trailer bogie of the vehicle. It was in this guise that the unit was used during by the rolling stock department on adhesion trails on the Shepperton branch from Strawberry Hill depot during October 1988 and again involving Class 442 units between Brookwood and Basingstoke the following April. It was sometime around 1990 that the formation of 053 was changed to the formation that ended up becoming the preserved unit. No one is quite sure when it happened, why or how. It is likely that the formation was changed at Strawberry Hill depot as both 053 and 054 were based there when not needed for engineering traffic. They had certainly been used during May and September 1989 as tractor units in formations with 1938 tube stock that were being taken from Strawberry Hill to Eastleigh Works for conversion for use on the Isle of Wight. What is believed to have happened shortly after this is that the Driving Trailer car from 054 was paired with the Driving Motor Brake from 053. Unit 054 continued in use with 053's driving trailer for a few years until it was reformed using another driving motor from the unit that had been 6289 to form a 'power twin' tractor unit.

The new look 053 turned up at Ashford Chart Leacon in August 1991 and underwent a general overhaul. One of the downsides of being departmental stock was that repairs and overhauls only focused on what was needed for their new role; thus the paintwork was not touched, nor the interior, just the traction and running gear along with the structure. Passenger doors were stripped of their exterior

The Preserved Unit • 151

Parked on the Fullwell Curve siding at Strawberry Hill depot in late 1995, unit 930 053 is still technically in service at this time but has not worked outside the depot in a while. It was at this time that it became the subject of preservation interest.
(Graeme Gleaves)

Parked in the company of several other withdrawn and stored EPB units, 930 053 is parked at Strawberry Hill whilst the decision on its future is being made and efforts to raise the funds to purchase it are frantically happening behind the scenes.
(Graeme Gleaves)

handles apart from four, one in each corner, needed for access or escape of test train staff. The DTS got the same treatment despite the fact that it still retained bulkheads in its compartments so staff could not move through the coach. In September 1991, 053 was fitted with the regulation high intensity headlight on the cab front at Fratton depot. The unit continued on tractor duties, including moving more tube stock from Eastleigh and the odd rescue of a defective unit and delivering it to either Strawberry Hill or Wimbledon Park. After its last repair at Selhurst in March 1993 the unit returned to Strawberry Hill and remained part of the tractor fleet under the auspices of the Engineering Development Unit based there. Along with the other units there it often spent long periods without any work and by mid 1995 was parked on the long siding adjacent to the depot having been out of use for some time. It was at this time that it came to my attention.

1995 was a year of transition on the Southern Region as it approached privatisation. The last of the EPB stock in service on the South Eastern Division suburban services was being withdrawn as they were replaced by new Class 465 and 466 Networker units. This prompted a movement to preserve examples of the stock that had been in use for over 40 years. This became The EPB Preservation group, of which I was a founder member, and very quickly the group had secured 2-EPB unit 6259, an example of a Southern built and operated British Railways unit. I had become aware of 930 053 parked out of use at Strawberry Hill and was also aware of its significance. There were a handful of ex-Tyneside vehicles still in departmental service but this was the last two-car in the correct formation with an unmolested Driving Trailer car. It represented part of a bigger story, the one you have read in the preceding chapters, and was the last chance to get a real Tyneside third rail passenger unit preserved, given that all the Met-Cam and NER passenger stock had been scrapped over thirty years before. I pitched the idea to the EPB Preservation Group who were not interested, deciding they wanted to focus on Southern Region stock. I believe this was the right decision for that group and they have gone on to do great things in their chosen field.

I did not want to pass this last opportunity up and in my job as a driver at Wimbledon Park I drove past the unit several times whilst it was parked on the siding adjacent to the Fullwell Curve. The only option was to go it alone and start a group to save it myself, which is exactly what I did in the spring of 1996. At the same time two coaches of the last surviving Class 503 unit became available and by combining the two projects I formed a group called The Mersey & Tyneside Electric Preservationists, or MATEP for short. Within very little time after pitching and networking the idea (in the days before social media) several others came on board and money started to be raised; there was just one problem, we had no idea who owned the unit and if it was for sale?

Investigations were made and proved complicated given that this was the transition period between British Rail and the privatised railway. All rolling stock was largely in the hands of three rolling stock leasing companies (ROSCOs) and none of them considered that they owned the unit. Likewise the freight and passenger operators, given that this was an engineering department unit, were approached and could not help. Eventually it was discovered that the unit had passed onto the books of the newly privatised infrastructure company, Railtrack. They had taken over all of the departmental stock that was used in clearing the infrastructure such as the de-icer fleet, and this unit, along with the power twin 054 had been included in the allocation.

Railtrack had inspected 053 during mid 1996 and decided it was surplus to requirements and put it up for disposal. The rescue bid was on. The bidding process was handled by the British Rail procurement department at Derby and during the summer of 1996 a price was agreed with them that was just above the scrap value and also included the cost of rail movement to Wimbledon Park, where it could be collected and transported to a new home by road. That left us with the need to arrange a new home. Whilst the bidding process was going on we had approached a fledgling new preservation site at Robertsbridge, The Rother Valley Railway (RVR), loosely connected to the Kent and East Sussex Railway, and asked if they were interested in housing the unit. It would be many years until they were able to run passenger trains and we wanted a base in the south of England as most of our volunteers were located there. The RVR offered siding space at their rudimentary initial site adjacent to Robertsbridge station car park. One lunchtime in late 1996, 930 053 was sandwiched between a two-car ex-SUB de-icer and tractor unit 014, formed of two ex-2-HAP motor coaches, and tripped the few miles from Strawberry Hill depot, via Kingston to Wimbledon Park depot and shunted into a siding. Unfortunately someone forgot to cut the unit out, which meant it was still connected to the third rail supply. The unit's emergency batteries, used for supplying emergency lights and feeding the control circuits, had not been topped up with water in years and whilst cut in they caught fire. Thankfully someone spotted this in time before any major damage was caused. The batteries themselves and the underframe box that carried them was all that got irreparably damaged. A close call but a sign that the next few weeks were not going to go smoothly.

The movement of the first car of the unit from Wimbledon to Robertsbridge was going to take place overnight on 1 December 1996 as it would be an abnormal load and needed to avoid the worst of London's traffic. The loading of the Driving Trailer car went smoothly enough and the next morning it was on the outskirts of Robertsbridge where heavy rain had fallen overnight. The lorry struggled to manoeuvre around the village streets and even pulled a telephone cable down that had been installed below the regulation height. Getting into the station car park proved too difficult an obstacle in the conditions and an additional tractor unit had to be dispatched to assist the lorry into place so it could eventually unload the car. After weeks of delay and a change of haulier the Driving Motor Brake car made the same journey, but this time with no drama, on 3 January 1997, and was successfully unloaded, despite there being snow on the ground. The unit was coupled up and put into its siding at Robertsbridge to begin the first part of its preservation story.

Restoration work began straight away and over the next couple of years a small team made progress renovating the interior of the semi-saloon behind the driving cab of the Driving Trailer car. The unit even had its compressed air system charged and brakes made operational so it could be shunted by the RVR's two diesel shunting locos during a couple of model railway exhibitions held at the village hall over the road from the station. We even got a small generator that could output at 70v DC and used it to test the lighting, which was found to be fully operational in the Driving Trailer but only partially working on the stripped out Driving Motor car. Thanks to units being sent for scrap

Now officially purchased for preservation, the last of the South Tyneside 2-EPB units sits at Wimbledon Park depot in November 1996 and is waiting for road transfer off the national network to begin a new life.
(Graeme Gleaves)

After a long and arduous journey the DTC sits on the road trailer at Robertsbridge and awaits final unloading at its new home. (Graeme Gleaves)

that had just the parts we needed we were able to obtain enough seating components to enable the Driving Motor to be fully rebuilt internally at a later date; these were stored in the empty Driving Motor Brake coach. MATEP changed its name to The Suburban Electric Railway Association (SERA) in 1998 as other stock had been acquired and located at the Coventry Steam Railway Centre.

All was going well, but then things went sour, as they have a habit of doing. The RVR had lost a lot of money trying to organise a mainline steam special that had failed to happen, and in an attempt to recoup their losses they asked all stock owners on the Robertsbridge

The DMBS car, ADB 977505, is rolled off the road trailer and onto Rother Valley Railway metals for the first time on 4 January 1997. (Graeme Gleaves)

On 4 January 1997 the two coaches of the South Tyneside EPB unit are coupled together in preservation for the first time at Robertsbridge. The snow on the roof of the DTC car nearest the camera is evidence that it had been delivered a few weeks prior. (Graeme Gleaves)

site for rent, something that had never been part of the original arrangement and something we, as a small group, could not afford. Whilst we had supported the RVR in the organising of fund raising events with them we felt that being asked to pay for their bad management was not in the spirit of our relationship. In the end, after threats of legal action by the RVR, a mediated agreement was reached but the relationship was too badly damaged and we sought to relocate our unit to Coventry to join other SERA stock there. We had no funds to move it and carried out an audacious publicity stunt to raise much needed funds by painting the Driving Motor coach in the black and white stripes of Newcastle United, complete with No. 9 and the legend SHEARER on the rear, and the logo of the shirt sponsors, Newcastle Brown Ale, on the cab front. It got its picture in every railway magazine in the country. We had an offer of some sponsorship from a company that ran a travel agency out of a station in County Durham if we put banner ads for them on our website, which we did, but the sponsorship money never materialised.

But as is often the case help can come from the most unlikely of sources. Railtrack was still continuing to operate an ageing fleet of EPB de-icing units and they were struggling to find spares to keep them going, so they approached us about 'borrowing' parts from our unit. We were able to negotiate the loan of these items provided they got the unit moved to Coventry. The move of both vehicles was effected during late September and October 1999.

156 • NORTH EASTERN ELECTRIC STOCK 1904–2020

The DTC car has been separated from its Driving Motor Brake partner and is positioned ready for collection at Robertsbridge.
(Graeme Gleaves)

Twenty-four hours after the previous image was taken the DTC car was safely at its new home at the Coventry Steam Railway Centre. It would remain here for nearly twenty years.
(Graeme Gleaves)

The Preserved Unit • 157

The controversial repaint of the DMBS car unveiled at Robertsbridge in preparation for the 1999 FA Cup final between Manchester United and Newcastle United. (Graeme Gleaves)

The rear end of the NUFC repaint showing the shirt number and name. (Graeme Gleaves)

We had a written undertaking from Railtrack that they could have the parts but were obliged to return them (or alternative equivalents) when no longer needed.

Little did we know that once coupled up and parked in a siding at its new home it would be seventeen years until any further meaningful restoration work would be carried out on the South Tyneside unit. The SERA had taken over the running of the Coventry Railway Centre in 2000 and the running of a site leaves precious little time or resource for rolling stock restoration, and all of the SERA fleet suffered from being out in the open with too few people to do far too many jobs. The site had started to attract the attention of vandals and other undesirables and windows got broken on several items of stock, including the South Tyneside EMU. At one point a vagrant was found sleeping in the partially restored semi-saloon; he had started a fire to keep warm which had burned a hole in the lino and floorboards and we had narrowly avoided having the whole coach gutted. In 2007 the SERA joined forces with a group of AC electric preservationists to form an organisation committed to providing a dedicated facility where the story of electric trains in the UK could be told. That organisation was The Electric Railway Museum Limited and during 2009 it became a registered charity and took over the running of the Coventry site. The SERA went into hiatus as all efforts were directed towards the charity getting the site ready to open to the public for the first time ever. One bit of unfinished business that was attended to during this period was the agreement with Railtrack for the return of parts borrowed several years earlier. Railtrack had morphed into Network Rail by this time and were disposing of the last of their EPB type stock. The SERA had obtained various parts from scrap units so ensured they had almost everything they needed to restore the Tyneside EPB with one exception, a correct express gear ratio power bogie. To Network Rail's credit they never questioned the original agreement paper work when it was presented to them and arranged for a power bogie to be given over to the project from scrap tractor unit 014 (one of the units that had delivered 930 053 from Strawberry Hill to Wimbledon Park in 1996). The SERA had to arrange the transport of it from the scrap yard in Yorkshire to Coventry which was done with little problem. The first Electric Railway Museum (ERM) open day was held during September 2010 and those first visitors got to see what all the visitors that followed during the next seven years saw as far as the South Tyneside EPB went, the unrestored unit sitting on the end of its siding waiting for its turn for work to recommence. It would take a dramatic turn of events for that to happen. Since shortly after the ERM opened for the first time, local business had designs on the land surrounding Coventry Airport, which included the ERM site. First there was the Local Enterprise Zone project that failed to get support as one of the shortlisted projects; that morphed into the Coventry Gateway scheme which would have seen multiple industrial units built in the area to form a massive trading estate. This went as far as to get planning permission but was called in and rejected by the Secretary of State. Finally in early 2017, just after ERM had signed a new three year lease on its site the Gateway project was relaunched to include a larger development that included a new engineering and test facility for Jaguar Land Rover on the opposite side of Rowley Road to the ERM. During 2017 the ERM was told their current lease was their last and they should make plans to vacate the land. With this news the SERA was re-activated and a work force assembled to promote the Tyneside project as it would need this if it was to find a new home.

Volunteers, some old hands and some new blood got themselves together as 'Team Tyneside' and began working on the unit for the first time since it arrived at Coventry; there was after all no point working on a site that was going to be razed in a matter of months. Discussions went on to find a new home and just before the final ERM open day in October 2017 an agreement was struck with the Battlefield Railway at Shackerstone in Leicestershire for the unit, the stores van that supports it and two industrial locos associated with it, one diesel and one electric, to go to the railway, where Team Tyneside would be an ongoing project. The movement of both cars from Coventry to their new home was carried out in April 2018. Team Tyneside had decided to complete the restoration of both the locos in its care first as these were close to fully operational. The South Tyneside unit was formed up and stored at Shackerstone and at the time of writing is next in line for a full restoration, when the team are free of all prior commitments, currently estimated to begin in late 2021 with the intention to take the unit back to its original condition as the only preserved North Eastern third rail passenger unit.

Pictured on 18 March 2018, the preserved unit is less than two months away from leaving the Electric Railway Museum for a new life at the Battlefield Railway. Work on the DMBS car is evident, along with the unseasonal snowfall. (Graeme Gleaves)

You can follow the project on social media:
Facebook: http://www.facebook.com/tynesideunit/
Twitter: @TeamTyneside

The Suburban Electric Railway Association website can be found at http://www.emus.co.uk

BIBLIOGRAPHY

Cooper, B. K., *Electric Trains in Britain* (Ian Allan Ltd, 1979)
Dunn, David, *British Railway Memories Tyneside Electrics: 1* (Book Law Publications, 2016)
Dunn, David, *British Railway Memories Tyneside Electrics: 2* (Book Law Publications, 2016)
Foster, Joan, *Newcastle upon Tyne A Pictorial History* (Phillimore & Co Ltd, 1995)
Hannah, Leslie, *Electricity Before Nationalisation* (The Johns Hopkins University Press, 1979)
Hoole, Kenneth, *The Electric Locomotives of the North Eastern Railway* (The Oakwood Press, 1988)
Hoole, Kenneth, *The North Eastern Electrics* (The Oakwood Press, 1987)
The London & North Eastern Railway Encyclopedia on www.lner.info
Moffat, Alistair & Rosie, George, *Tyneside, A History of Newcastle and Gateshead From Earliest Times* (Mainstream Publishing, 2006)
Sprague, Frank J., The Multiple Unit System For Electric Railways (*Cassier's Magazine*, 1899)
Stobbs, Allan W., *Memories of the LNER Tyneside* (Allan W. Stobbs M.S., B.Sc., 1988)
Two-Car Electric Suburban Units (*The Railway Gazette*, April 1955)
Tyne & Wear Metro Fact Sheets (Tyne & Wear Passenger Transport Executive, 1989–1982)
Young, Alan, *Suburban Railways of Tyneside* (Martin Bairstow, 1999)
Locomotives of the LNER (Railway Correspondence & Travel Society, 1990)

INDEX

Accidents involving railways, 39, 47, 72-3, 91, 140

Beeching, Dr. Richard, 92-3, 97
Blyth and Tyne Railway, 19, 115, 118
British Thompson Houston, 26, 43, 131
British Railways,
    formation, 67
    new electric stock built by, 79, 86
    renumbering of vehicles & locomotives, 69-70, 89, 134, 142, 145
    stations closed by, 75, 91, 108
Brush Electrical Engineering, 131
Buckeye couplers, 81, 87
Butterworth, Alexander, 135, 137
Byker viaduct, 111

Carville power station, 25
Coal mining, 9-11, 48
Conductor rails, 24-5, 61-2, 96, 106-7
Controlled set, 33, 38, 47, 60-1, 91
Cowhead couplers, 30, 56, 60
Crompton Parkinson, 56

Darlington works, 139, 142-3, 145
Diesel multiple units, 93, 99, 106, 108

East coast main line electrification plans, 142-3, 145
Eastleigh works, 79, 84, 87, 96, 147, 149
Electric locomotives,
    liveries, 131, 134, 139, 143
    scrapping of, 135, 140, 142, 145
    types of motor equipment fitted, 131, 139, 143
Electric multiple units,
    de-icing van conversions, 35, 62, 98
    liveries, 29, 43, 47, 51, 61, 64, 120, 123, 126-8, 149
    motor luggage vans, 30-1, 42-3, 60, 86, 97, 99
    numbering of, 31, 43, 60, 84, 120, 126
    pram van conversions, 71, 74, 93
    renumbering of, 68-70, 72-3, 92, 149-50
    scrapping of, 62, 70, 72, 84, 97, 101-2
    types of motor equipment fitted, 30-1, 43, 56, 80, 119
Electro pneumatic brake, 56-7, 60, 81
English Electric, 80, 140

Forth Banks power station, 14

Gibb, George, 21, 30
Gloucester Railway Carriage & Wagon Company, 50
Great Northern Railway, 45
Gresley, Nigel, 56, 140, 144-5

Heaton car sheds, 30, 46, 132
    fire at & aftermath, 39, 41-2
High Level Bridge, 52, 71, 83-4, 93-4, 96
    construction & opening, 10, 49
Hudson, George, 18

Ilford depot, 142

Lancashire & Yorkshire Railway, 27, 36
London & North Eastern Railway,
    formation, 45
    electrification plans, 48-9
    electric multiple units, 56-60, 64-5
    renumbering of vehicles & locomotives, 47, 134, 142, 145
    replacement by British Railways, 67
    stations opened by, 48, 66
London & South Western Railway, 23

McLellan, William, 14, 23
Metrocars,
    construction, 115, 119-20
    electrical equipment, 120
    refurbishment, 126
Metropolitan-Cammell (Met-Cam), 50, 56, 115
Metropolitan Vickers, 49, 143
Metropolitan West Side Elevated Railroad, 24
Mertz, Charles, 14-5, 23-4, 129, 135, 145
Mertz, John, 14
Mertz & McLellan, 14, 21, 23, 25-6, 135, 142

National Railway Museum, 120, 135-6
Neptune Bank power station, 14
New Bridge Street station, 19, 36
    electrification of, 27

Newcastle & North Shields Railway, 18-9, 127
Newcastle Central station, 20, 36, 38, 52, 99, 106
   accidents at, 72, 91
   construction & opening, 10, 13, 15
   diamond crossing, 16, 33
   metro station, 109
Newcastle-upon-Tyne Electric Supply Company, 14, 23, 25-6
Newport to Shildon route, 137-9, 143
North British Railway, 45
North Eastern Railway,
   formation of, 19
   competition with trams, 15, 21, 31
   electrification plans, 23-26, 36, 49, 130-1, 135, 137-138, 142
   electric multiple units, 27-35, 43-4
   locomotives, 130-1, 139, 142-144
Northern Rail, 126

Overhead wires, 107, 115, 131, 138, 142, 145

Parsons, Charles, 12, 14, 25
Pardon Dene power station, 14
Ponteland branch line, 26, 39, 108

Quayside branch, 36, 130-2, 142
Queen Elizabeth II, 113, 122, 126
Queen Victoria, 10, 15, 49

Raven, Vincent, 30, 37, 137-8, 145
   early career & appointment as CME, 24, 130, 135
   electric locomotive design, 139, 142
   retirement, 144
Riverside branch,
   route and opening, 19
   electrification, 26, 36
   operations, 38, 62, 74, 91
   closure, 97, 106-7

Siemens Brothers & Co, 26, 139
South Gosforth car sheds, 62, 78, 84, 121-2, 135, 142
   construction and opening, 42, 45
   conversion for Metro use, 113, 115, 120
South Shields,
   first railways, 49
   electrification of route to, 52-3
   de-electrification of route, 96, 106, 108
   services to, 91, 93-4
   metro services to, 107, 112-4
Sprague, Frank Julian, 24, 39
Substations, 26, 49, 54, 106-8, 138

Thompson, Edward, 145
Tyne & Wear Passenger Transport Executive, 107, 119, 126
Tyne & Wear Metro,
   construction, 107-9, 112-3, 126
   rebuilding of former BR stations, 111-4
   test track, 115

Westinghouse
   air brake, 30, 32, 56
   electrical equipment, 57, 80
Westinghouse, George, 26
Wordsell, Wilson, 23, 30, 130, 135

York, Newcastle & Berwick Railway, 18
York carriage works, 27, 30-1, 42, 45, 50-1